Human or F...,.....
People and Places of Edge Hill

Edited by Sylvia Woodhead and Ann Chapman

Green Lane Books in association with

accredited by Lancaster University

First published in Great Britain by Green Lane Books in 2006
Green Lane Books
Oakleigh, Back Road,
Lindale, Grange-over-Sands,
LA11 6LQ

e-mail: sales@greenlanebooks.co.uk

ISBN 1 900230 41 0

Designed and produced by Green Lane Books
Cover design by Andy Butler
Printed by The Printroom (UK) Ltd
16 Crosby Road North
Waterloo
Liverpool, L22 0NY

Contents

List of Figures

7: The Days of Diversification

8: The Thatcher Years: the 1980s

9: Towards the End of the Century

10: Practice makes Perfect

11: Geography United

Postscript

List of Contributors

Rachel Bowles, Honorary Research Associate, School of Education, University of Greenwich

John Cater, Director and Chief Executive, Edge Hill

Ann Chapman, Cartographer, Edge Hill

Kathryn Coffey, Geography Technician, Edge Hill

Peter Cundill, Senior Lecturer in Geography, University of St Andrews

James Curwen, Psychology Technician, Edge Hill

June Ennis, Geography Administrator, Edge Hill

Andrew Griffiths, Director of Wessex Trains

Jamie Halsall, Postgraduate Geography Research Student, Edge Hill

Vanessa Holden, Postgraduate Geography Research Student, Edge Hill

Neil Immins, formerly Senior Lecturer in Geography and Education, Edge Hill

Nick James, The Open University

Barbara Lang, Postdoctoral Researcher in Earth Sciences, Liverpool University

Fiona Lewis, Freelance Writer, formerly Lecturer in Geography, Edge Hill

Gerry Lucas, Senior Lecturer in Geography, Edge Hill

Bill Marsden, Emeritus Professor of Education, University of Liverpool

Gregg Paget, Senior Lecturer, Department of Environmental and Geographical Sciences, Manchester Metropolitan University

Mike Pearson, formerly Geography Co-ordinator, Edge Hill Woodlands campus

Colin Pickthall, former MP for West Lancashire

Ann Power, Postgraduate Geography Research Student, Edge Hill

Charles Rawding, Geography PGCE course leader, Edge Hill

Nigel Richardson, Head of NGAS, Edge Hill

Paul Rodaway, Director of the Centre for Learning and Teaching, University of Paisley

Ann Power, Postgraduate Geography Research Student, Edge Hill

Tasleem Shakur, Director of ICDES, Edge Hill

Nigel Simons, Associate Dean, Faculty of HMSAS, Edge Hill

Peter Stein, GIS Technical Support Officer, Edge Hill

Stephen Suggitt, Senior Lecturer in Physical Geography, Edge Hill

Joan Swinhoe, Former Geography Administrator, Edge Hill

Sylvia Woodhead, Senior Lecturer in Geography, Edge Hill

Ann Worsley, Senior Lecturer in Physical Geography, Edge Hill

Acknowledgements

This volume is produced with support from Edge Hill's Teaching and Learning Development Unit and HMSAS Faculty and the Manchester Geographical Society.

Thanks are due to all the many people who supplied information and answered questions, John Alcock of Slack House Farm, Adrian Allan, Liverpool University Archivist, Pauline Bankes, Ian Bond, Conservation Officer of West Lancashire District Council, Mark Burton of Estates, John Chapman for patiently searching through Territorial Army minutes, Kate Chapman and Daniel Tetlaw for photographing Bingley College, Mona Duggan, John Entwhistle, Mark Gaskell for wartime records, Steve Igoe, David Lambert and Frances Soar at the Geographical Association, Joe McNamara, Mrs Nanson, Ormskirk Historical Society, staff in the library who helped archive searches. Thanks also to Roy Bayfield in Marketing, to Sharon Buckley, Alumni Officer, for sending out questionnaires to members of the former students' Guild, and to Roisin Rowley-Smith for drafting press articles calling for wartime memories.

The book could not have been produced without the contributions of colleagues past and present, and especial thanks go to all those former students (from 1935-1964) who sent in their memories of studying geography at Edge Hill. It has been a privilege to share their reminiscences. Current geography students have also helped by completing questionnaires about the buildings. Many thanks are due to Paul Rodaway and Michael Chapman for their useful advice on earlier drafts, and to our families for their encouragement, love and support.

While every effort has been made to ensure the accuracy of the contents, responsibility for any errors that have made it through to the final version is entirely ours.

Sylvia Woodhead
Ann Chapman
Edge Hill November 2005

Foreword

Last summer Sylvia Woodhead and Ann Chapman appeared at the office door. Fifteen minutes later I'd agreed to leave HE administration and politics for the first time in a decade to reflect on what geography meant to me. Now, as I read the contents list, I know just how persuasive Sylvia and Ann must be. Edge Hill people from the 1930s to the present day grabbed their pens, hit the keys and, under Sylvia's direction, have produced a wonderfully varied sense of both Geography and this excellent institution.

My love affair with Geography has always been fuelled by the subject's eclecticism. From lentils to post-modernism, Norway to Bangladesh, the blitz to e-geography, little is beyond the geographer's reach. There is no better undergraduate discipline for understanding the worlds we live in and the lives we lead. No other subject embraces such a range of partner disciplines, from the physical sciences to the social sciences and across the humanities to the arts. No other subject develops such a range of key skills – numeracy, literacy, artistic representation through diagrams and maps, fieldwork, presentation skills, IT skills, a sense of place, a sense of space. And no other subject offers so much capacity for fun along the way.

But life isn't easy for Geography. Squeezed by an increasingly instrumental school curriculum in the nineties, pressured by the demand for overt vocationalism in the noughties. Suffering for its very strength – a lack of direct cognacy with a single set of disciplines as the post-14 student specialises in the 'sciences' or the 'arts'.

And, even in higher education, there are challenging times to come, with few commentators confident of what the advent of a market system will mean for traditional subject choices. But, as a book like this demonstrates there is no subject more vital, more alive, and there are many committed geographers dedicated to the discipline's future health.

This book is an equally eclectic life story of Edge Hill. From birth in inner Liverpool, following a Betjeman-like route beyond suburbia in the 1930s, through difficult wartime days in Bingley, co-education and the sixties, the loss of many peers ten years after, and the departing local authority parent in 1989, and all told through the eyes and words of participant observers. Looking back, perhaps the 170 teacher training colleges of the early 1970s weren't meant to survive as independent entities, but a few of us are still here today. And we are much stronger and more diverse than ever before. To origins in teacher education have been added parallel strengths in the field of health, and the original subject elements of programmes, going right back to 1892, laid the foundation for the 360+ degree routes we see today.

As I write, an institution of over 13,000 students and 1,440 staff, with a turnover of £50m a year, is in the process of applying for taught degree awarding powers and University title. Over £60m has been invested in new infrastructure in the past decade, applications are at record levels (up 62% in five years) and demand for 2005/2006 is surpassing all previous records. Nonetheless Edge Hill, like Geography, has an interesting, even challenging, few years ahead. But, with a government and a market committed to recognising excellence in teaching, and with geographers and subject teams offering high quality learning opportunities properly underpinned by research and scholarship, those interesting times can be hugely successful times too.

My thanks to all the contributors who have made this collection possible. Enjoy the book, and enjoy the future challenges too.

Dr. John Cater
Director and Chief Executive

Introduction

Are old buildings important? Amazingly even modern 2005 students, those sophisticated image-conscious people, think they are. This volume is based on the premise that buildings have an important effect on the lives of the people who live and work in them. Many of its authors currently occupy, or once occupied, an old wartime hut which alternately exasperates and entrances them. Built in the Second World War, the building is plainly not suited to twenty-first century academic life; the rooms are the wrong size and shapes and its corridors are too small. It ought to be demolished, and yet no one really advocates this.

The building, a former Army block, used now for teaching and research in geography, holds sixty years of memories of people who worked and lived in it, and many of these memories have been captured and presented in this book. Investigations have gone right back to the very founding of Edge Hill in 1885 to explore the relationships between the people, staff and students, and the places they lived and worked in. Edge Hill's campus has been on three sites over time, Liverpool, Ormskirk and Bingley; three places to explore, three different environments where students and staff have interrelated with each other and their surroundings. This book aims to tell the stories and geographies of these sites, and to demonstrate the distinctiveness of geography at Edge Hill, now and in the past. We review geography teaching and research, from its earliest beginnings to the present day, and aim to show where geography at Edge Hill started and how it has developed over the years. Past and present geographers tell of their time and contributions, and explain the significance of their research. For earlier years the story is taken up by past students, who recount their memories of learning geography at Edge Hill, in the amazing archive of the Guild of former students' newsletters from before 1900 to present, and through personal contacts. This is a geographical perspective, a personal investigation of a single discipline followed through the history of a College of Higher Education, with the recurring theme of people and places, attempting to reconstruct how staff and students have reacted to the places, Liverpool, Ormskirk and Bingley in which they were studying. It also builds on the previous history of Edge Hill written by Fiona Montgomery in 1997, though drawing on a series of essays by more recent students and staff.

When founded in 1885 as a Teacher Training College for women in the Edge Hill district of Liverpool, from which it derived its name, geography was one of the first compulsory subjects to be included in the syllabus. In 1933, the College transferred to the outskirts of Ormskirk to its present buildings, which had to be evacuated in the war for use as a military hospital. From 1939 to 1946 cohorts of Edge Hill students were taught at Bingley in West Yorkshire. It was during the war when the main College block was used as a military hospital that the present geography building was constructed, as 'Block A'. Some students returning to the Ormskirk campus in 1946 were accommodated in the building which after being painted white, was termed 'the White House'. It formed accommodation for the first intake of male students in 1960, and became an education block when students transferred to the newly constructed halls of residence. When a new education block was built, geography moved into its existing building.

The story that follows recounts the people and their heritage of memories about their geography education at Edge Hill. A chronological approach is taken, of College sites, their surroundings, the staff and students and their perceptions of the teaching of geography at that time. This is linked to selected developments in society and technology, and to differing approaches in the discipline of geography. It is intended to show how geography developed and also to try to recreate a sense of place, those feelings the staff and students developed for the locations in which they worked.

In the Prologue two Edge Hill undergraduates, now research students, explain the sense of place they felt at Edge Hill. **Vanessa Holden** writes about why she liked Edge Hill as an undergraduate, and later explains her research and love of the coast. **Jamie Halsall** reflects on what it was about (human) geography teaching at Edge Hill that made him switch from his original focus on Information Technology. He ponders on the close Edge Hill – Liverpool link, which the rest of the book develops further.

Empire Days

Edge Hill College began in 1885 in the Liverpool district of Edge Hill in a large house in Durning Road, built by George Holt, who had links with Liverpool's shipping past. Although adapted for teaching, with classrooms and dormitories in an additional wing, accommodation was always crowded, and by modern standards very primitive. Photographs show stiffly formal young women, attired in hats and long dresses. Yet despite this there were picnics on the Wirral, and an annual visit to the Walker Art Gallery. Before the turn of the century, University students took a three-year course and examinations and some tuition, in French and History, in the then University College of Liverpool. The geography staff, all women as were the students, showed lantern slides, and led field visits: Miss Deakin by public transport to the Wirral to see streams, rocks and submerged forests on the beach, while Miss Butterworth, later College Principal, wrote textbooks, which reflected the dominant colonial views of the time, and led coach tour geography excursions to the Lake District. Fieldwork started early at Edge Hill, and although its mode of transport and destinations have changed, it remains important today. Edge Hill was a pioneer in providing a non-denominational geographical education for women, who later became schoolteachers in classrooms walled with posters of imperial propaganda. Evidence suggests that methods and resources were at least as good if not better than the norm for the time, when education was largely limited to the privileged male few. In 1925 Edge Hill College was acquired by Lancashire Education Committee who constructed an entirely new campus on a greenfield site just outside Ormskirk, a small market town on the West Lancashire Plain.

Ormskirk before Edge Hill

Ormskirk's history is poorly documented, as the place was always off the beaten track and avoided by early travellers to Lancashire who feared getting lost, or drowned, in the surrounding mosslands. Famous as a market town, Ormskirk's market has been its most enduring characteristic. Its fortunes have varied with those of the Grade 1 potato producing agricultural area surrounding it, much as now. Many enterprises, successful elsewhere, failed in Ormskirk. Silk and cotton mills, soap works, rope and linen works have all started and closed. There were even attempts to develop Ormskirk as a spa, modelled on Tunbridge Wells, but in all cases it proved too remote to attract visitors and lost trade and clientele to other areas. Famous also for the patronage of the Earls of Derby, the landowning Stanley family who lived at one time at the nearby Lathom House, Ormskirk lost much trade when the family moved to their new home in Knowsley, Liverpool, and moved their races to the more famous Aintree. Ormskirk has long been overshadowed in its development by Liverpool, Southport and Wigan.

This Green Campus

Edge Hill students transferring from city life in Liverpool to the countryside site adjacent to Ormskirk must have experienced a severe culture shock. Their educational experience changed; it was more difficult to attract evening and weekend speakers, excursions were more difficult to organise and links with visits to the Wirral were broken. However, the Ormskirk site provided brand new purpose built spacious modern accommodation and space for playing fields. Edge Hill's green campus continues to be a major attraction to present day students. The extensive grounds were lovingly developed with flower beds, rose gardens, trees and a kitchen garden which provided fresh produce to supplement the diet; girls were expected to pick fruit and vegetables in season and help with some domestic chores. Students had to accustom themselves to views of the rockeries, rows of cabbages in fields and the water tower, instead of the terraced houses of Edge Hill in Liverpool. **Eileen Boocock** and **Irene Mackenzie** share their memories of the comprehensive pre-war geography course, taught by Miss Butterworth, in the Education Block in the main building.

The Campus at War

After only six years in Ormskirk, Edge Hill College had to move again as the War Office requisitioned the site in 1939 as a military hospital, and, before the evacuation of staff and students was completed, a new set of Army people, Royal Army Medical Corps, Territorial Army and Queen Alexandra's Nurses, moved in. **Peter Stein** recounts some results of his investigations into preparations for the war. The main College building operated as a hospital, and its corridors and lifts were adapted to accommodate hospital trolleys, and all the internal doors were taken off. Both Army staff and patients were there only for a short time, but **David Owen** nevertheless has clear memories of training at Edge Hill for later active service abroad. Great secrecy surrounded the operation of the military hospital and few were allowed in; certainly **Grace Hardisty** when driving an ambulance up to the front door hardly set foot in the hospital. War office records detail the illnesses of sick soldiers invalided from wartime service abroad who were treated at the military hospital.

It was sometime during the war that the present geography building was constructed as Block A, probably just as an additional hospital ward, possibly an isolation ward for TB patients, though why it was constructed in brick at the front of the College, while Block B onwards, including an Officer's Mess and additional wards, were built behind the main College buildings, remains an intriguing question. Evidence for wartime activity is sketchy, but the Ormskirk area did sustain hits and the campus possibly had a bomb crater at this time. The importance of wartime heritage is being more widely recognised, with both the BBC and Liverpool Museums having staged big campaigns to collect memories. A little more is known of the Saturday evening dances at the military hospital, which led to many romances. **Beryl Reeve** writes of encountering her future husband at one of these dances. The Polish Army occupied the hospital for the final months of the war, leaving more unsolved mysteries: about how many soldiers died, and who might be the ghost of the geography building.

The Bingley Years

Meanwhile the whole College had moved to Bingley in West Yorkshire to share accommodation with Bingley College. I doubt that either College was pleased with the arrangement, but the best was made of the situation on both sides. Edge Hill retained its name and independence, the Colleges simply doubled up. All bedrooms were shared, and the two different timetables were dovetailed together. Wartime conditions were hard; food was rationed, winters were severe, blackout had to be maintained. Despite these difficulties, education, including that in geography, still taught by Miss Butterworth, continued in a different, more hilly environment, with field visits to the limestone scenery of the Yorkshire Dales (not recognised as a National Park until 1956) to Malham, Ingleborough Cave and Gaping Gill. Miss Butterworth became Principal, and Miss Dora Smith took over the geography teaching. **Constance Bancroft** and **Alice Dale** write about their very clear memories of geography excursions, while **Elma Eastwood** remembers geographical studies near her home. It is strange to think that for six years from 1939 to 1946 Edge Hill students did not know the Ormskirk campus. To this day they refer to themselves as 'Bingleyites', and have clear memories of frequent Air Raid practices disrupting their learning.

Post War Austerity

The move back to the Ormskirk campus in 1946 must have been equally traumatic, although **Dorothy Firth** recounts how excited the students were to return. The second year students did not know their way around, and so could not take the normal Edge Hill role of 'sister' (or 'mother') to first years; they would have learnt their way round together. The site must have looked very wild; much of the land had been ploughed up for the 'Dig for Victory' campaign during the war, and the gardens had not been maintained, so the rock garden in particular was a mass of weeds. The present car park in front of the geography building had also been dug up and cultivated. Miss Butterworth, now the Principal, put much effort into renewing the grounds, enlisting students for example to clear up the enormous number of dandelions which had grown, as **Gladys Cooke** describes in her memories of 'the good days'. Planning the continued development of the grounds was a recurrent theme of the Principal, Gardening lecturers, students, and gardeners, with a literal love affair in the gardens for Derek and **Sue Sumner**. Edge Hill had a very active Rural Studies department up to the 1970s.

Miss Butterworth was apparently horrified at the Army hut which had been put up near the front entrance. It is interesting that she immediately had it painted white, and didn't order its demolition. Material goods were in short supply in the post war years, and she had the hut adapted for students to live in; **Olive McComb** recounts her two years in the 1940s in the primitive, but well guarded adapted hall of residence. Geography teaching continued in the main block, at first downstairs where the current Directorate offices are, then later upstairs, at the front of the building. In the late 1940s and early 1950s conditions generally were quite difficult, with few textbooks, poor quality paper, rationed food, little money and Victorian strictness still maintained in the College. **Lois Ford** has kept her geography work, exam papers and College Regulations, with exam grades and results for the whole geography class, and a book she bound in an extra book binding course she took at Edge Hill, after taking her finals, and before being allowed to leave. Other students have only odd idiosyncratic memories of geography lectures and lecturers, though reminiscences of field 'expeditions' are usually strong. The view of the water tower figures in recollections of studying geography at Edge Hill, perhaps the most potent landscape symbol in the area.

Gradually the College settled down after the war, to continue the two-year teacher-training course, and many Edge Hill students went into the world to a highly successful career in teaching; one student for example became the first female head of a school in Scotland. In the 1950s a clear group of geographers began to emerge from the records. The head of department from 1958 onwards was Miss Anna Cooper, who, as a historical geographer, reflected the continuing theme in geography that geographers cannot make sense of the world unless they know the origins of their subject. Like the rest of the staff at that time, she was a former teacher as all the courses were education based. **Dorothy Collings** and **Margaret Catterall** write about coping with the harsh living conditions in the 'White House' approaching the 1960s.

Back to the Future

The 1960s saw many changes; the first male students were recruited to Edge Hill, and were accommodated at first in the geography building; the White House. Other new halls of residence were constructed in modernist style and the White House next housed the Infant Studies for the education department, the Audio Visual unit, before becoming the geography building as it is today. **Jean Cowgill** writes about her memories of Edge Hill as being a 'little bit of Wales in Lancashire', as Mr Williams and Miss Williams taught geography, and students went on fieldwork to Wales to help Mr Williams with his research on the coast near Barmouth. The 1960s saw the re-introduction of three-year University courses, and more staff. **Rachel Bowles** recounts how closeness to a decent choir (Liverpool Philharmonic) formed part of her decision to move to Edge Hill in 1965, to teach geomorphology on the new BEd degrees, and how she felt at home with the geography team. She found that quantitative geography and problem solving methods made it a good time to be innovative. Fieldwork was important, reflecting the diversity of nearby landscapes, and the team ensured that it held prime place in the timetable. Reflecting on changes then and now, she notes the division, which concerns current geographers, between academic geography and its concepts, and school teaching applications. She worries about the geographical inexperience of modern students who go to their local University and study a fragmented modular syllabus.

    Bill Marsden, appointed the following year, when the number of geographers had risen to six, remembers that the Edge Hill geography department gave a useful boost to his later career in geographical education, a rite of passage between school teaching and the Liverpool University Education Department. He remembers field trips to Arnside and Whitby, which still continue today, the expertise in historical geography and the extensive map collection, both still maintained. He recollects Edge Hill at a transition stage from an all female teacher training College to a mixed HE institution. He enjoyed working with colleagues, and still remains in touch with some. Vic Keyte, a human geographer of great character and significance to geography at Edge Hill began in the mid 1960s expansion, while Ann Smith was a gifted historical geographer who later became Dean of the School of Humanities.

Days of Diversification

The 1970s saw the first new BA degrees, including geography, and brought further new staff. **Peter Cundill**, a geomorphologist, now at

St Andrews University in Scotland, was the first geography appointment with a PhD He writes about his experiences at Edge Hill during this exciting time of change, as the departmental focus changed from education to the academic study of geography underpinned by staff research. Initially he taught, mainly as a biogeographer, on education degrees, and had to supervise student teaching practice, while helping to plan the new Lancaster University degree courses in physical geography. Students did surveying, and analysis of sand and soil samples in their practical work and local fieldwork, and a residential field course took place in North Wales. He explains how his research on grains of pollen in sediments can help discover the development of past landscapes and vegetation. Peter left Edge Hill before he delivered the modules he developed. These were taught by his replacement Julia Franklin, later McMorrow, who, like Nigel Simons also appointed in the 1970s, had a research background. Before he left Edge Hill Peter had an unnerving experience one evening, though he never actually saw the ghost. Geography in the 1970s was not all work: he was one of the geographers who contributed to the Edge Hill staff cricket team.

Despite switching to Lancaster University degrees, links between geography at Edge Hill and Liverpool University remained strong. David Halsall, a research student at Liverpool University when Peter Cundill was a tutor there, was appointed to Edge Hill as a human geographer in the year after Peter's appointment. David had research interests in transport and taught industrial geography and gender issues. Andrew Francis later in the 1970s found geography full of friendly people, though he has since returned to his first love of planning. **Neil Immins**, appointed in 1975 to teach human geography, and curriculum geography on BEd degrees for prospective teacher, has pleasurable memories of supervising students on teaching practice. Neil reflects on the relaxed days of the 1970s, when staff had time to meet and engage in social activities, and a good performance in cricket was expected. He remembers with affection his former colleague Paul Gamble who sadly died at an early age. **Colin Pickthall**, a lecturer at Edge Hill from 1970 to 1992 in English then European Studies, remembers geography colleagues and their strong commitments to his dramatic productions of Shakespeare and Brecht. He assesses them for their ability to enunciate, remember lines and move on stage - surely key skills of a geographer.

The Thatcher Years: 1980s

John Cater starts *A Personal Geography* at primary school, as a pupil enamoured with maps. He describes being initially beaten at interview for a post lecturing in geography at Edge Hill by Andrew Francis, and the steep learning curve required of a new lecturer. He recounts the hard work behind the publication of *A Social Geography*, which he co-authored. The text, a 'real-world' look at social issues in work, housing, crime and disorder, was a best seller, selling 23,000 copies. He discusses his belief in the importance of structures in society, a structuralist approach to geography. **Gregg Paget** reflects on Edge Hill's attractive campus, helpful lecturers and friendly students. While living at Southport he experienced the very cold winter of 1981-1982 when parts of the marine lake froze. He has fond memories of field trips, of cycling in the countryside, and the attractions of Liverpool, both social and for academic interest in urban planning. He found Edge Hill a small friendly place, where staff got a lot out of students. **Mike Pearson,** in an input from one of Edge Hill's 'edges', writes about his almost one-man-band act at the Woodlands campus at Chorley, where for many years he devised and led geographical courses and conferences for teachers. Geography in the 1980s and 1990s was probably the least understood subject for primary teachers, who in many cases had dropped the subject early in their own schooling. He describes how school inspectors, as now, were worried that geography was not being taught well. Mike's courses boosted teachers' geographical knowledge, many of whom now occupy eminent positions. Mike describes the amazing web of geographical contacts he built up in the North West, his work as a Regional Co-coordinator for the Geographical Association, and the recognition he earned for his many achievements. **Nigel Simons** writes on policy and practice in teaching human geography; he reflects on whether geography's subject diversity is a weakness, as in the human-physical split evident in some years, or strength, as in the recurring environmental theme of much of Edge Hill's geography. Supporting the latter view he argues that resource management and coastal zone management have been popular themes that still continue, albeit with different staff, in Edge Hill geography today. Ironically after taking these strengths to Preston, Nigel has now brought his geographical experience back to Edge Hill. **Joan Swinhoe**, officially a technician, but really clerical support for a staff of nine people, who, apart from the Anna Cooper and Julia McMorrow, were all male, did typing for staff, booked coaches and fielded student questions. Some of the older members of staff were perhaps a little more intimidating than in modern times: Joan also offers little pen pictures of some of her former colleagues.

Towards the End of the Century

Andrew Griffiths notes his venture northwards as a lecturer in human geography from Exeter University following his research on farm diversification. Never comfortable with lecturing, he did not complete his PhD, and decided that the northern academic life was not for him, and returned to Exeter University to run the Tourism Research Group. Recently he has taken up an interest of which David Halsall would approve: Andrew is now manager for Wessex trains, and judging by the picture he has sent of himself he is thoroughly enjoying no longer having essays to mark. Derek Mottershead became head of geography in the early 1990s, replacing Anna Cooper who had led the geography team for so long. Derek, a geomorphologist with research interests in rock weathering, was also co-author of an undergraduate geography text. He upgraded the geography building, making better laboratories and using the wartime spaces more effectively. **Fiona Lewis** remembers Edge Hill as a nice place to work, with the geography building away from the main block, in its own space.

Paul Rodaway, a very modern, in fact a post-modern geographer as he describes himself, was another transfer from 'down south', from University College, Chichester. Paul brought a fresh and vibrant approach as a human geographer, questioning approaches, and interested in how people learn about and react to places; a social and cultural geography. On appointment at Edge Hill he had just completed a ground breaking text on *Sensuous Geographies: Body, Sense and Place*, in which he looked at the geography of the human body and its senses, and then considered the post-modern 'sensual' experiences in shopping malls or in TV soaps. Paul also explored, in his teaching and research, alternative geographies, of disabilities and other cultures, and virtual geographies of cyberspace. Supportive of fieldwork, Paul describes encounters with many different places, Cuba, China and Morocco. He remembers dissertation supervision, on topics like *The Simpsons* and spatiality in video games, like *Final Fantasy X*. Now funnelling his enthusiasm into learning and teaching at the University of Paisley, Paul remains at heart a geographer.

Paul Rodaway joined geography at Edge Hill at the same time as Kath Sambles, a hydrologist, a physical geographer, who was interested in instrumentation to measure snowfall. Kath readily embraced new technology and developed, with Gerry Lucas, a student friendly computer programme, Geo-Maths, to help students with geographical uses of maths. Like most geographers Kath liked the outdoors and worked as a voluntary Ranger at a local Country Park, where she developed an Art and Education Trail through the woodland. Kath had her own romance, meeting her future husband on a walking holiday; the downside was that he lived in Australia, so she emigrated to be with him.

Paul Rodaway brought James Newman, a research student, to Edge Hill and was supervisor for his research into the world of video games. James always felt 'on the edge' of Edge Hill, as some regarded his PhD topic as being only marginally geography, and his main contribution was to teach on the first year field course to Glasgow, as did Richard Jones, another research student. Richard was looking at early Holocene change preserved in the carbonate sediments at Hawes Water, which provided an excellent climate record. Richard's own romance was with Sharon Gedye, soon to be lecturer in physical geography at Edge Hill. She claimed to be one of the first 'Sharons' in the country to gain a PhD From her research she developed a second year module called Peatlands, which attracted a small but keen cohort of students. Sharon developed WebCT delivery of modules. It was a natural extension of this to move back to her native Cornwall to the Geography Subject centre. Sharon was very proud of her unusual Cornish surname, and was a little surprised, on the first year geography field course to south Cumbria, to encounter an enclave of Gedyes, as estate agents and solicitors, at Grange-over-Sands. They were, of course, related.

In the 1990s geography had a series of PhD students, including Fiona Mann and Gareth Thompson who were researching climate change. **Nick James** was loosely based at Edge Hill for his research jointly supported by Afro-Asian Studies and Geography. Nick's research was on food security in the Gokwe region of northern Zimbabwe, a politically unstable area, with poor soil and tsetse flies where many thousands of people who had been forcibly relocated had no links or initial understanding of the soil characteristics or the names of the plants - they renamed trees according to what they recognised from their home region. Nick gained the support and respect of local people by his mastery of their language, and his patient questioning of their environmental knowledge. His account takes a series of leaps from sweeping up autumn leaves to the greenhouse effect, food and cotton. He concludes that cotton cropping, though important to Zimbabwe's economy, is impoverishing both soil and people. His own life has been intricately linked with Africa, not least from calling his fantasy football team the Zimba boys after the village where he grew up. **Barbara Lang** recounts a personal story, painful and inspiring, of her transformation from housewife, mother and business partner, to postgraduate pushing back the boundaries of understanding of past climate change. Using cores of sediments drilled from a small lake near Silverdale, she extracted microscopic heads of midges, in a technique she had to devise. Her results showed that the midges responded to cooling events and this small lake was sensitive to temperature changes occurring in the North Atlantic Ocean. This could be a benchmark for the future: her research helps to understand how ocean circulation affects climate inland.

Practice makes Perfect

Since 2000, geography at Edge Hill has had largely its present complement of staff. Sabbaticals, when staff negotiate a term or even a year off to carry out their research, have proved difficult to organise, creating more stability in personnel. **Kathryn Coffey** ponders on the changing nature of work as a geography technician, providing technical support for staff and students in the lab and field, and maintaining equipment to analyse water and sediment samples, and making thin sections of rocks. She describes the pleasures of seeing students grow in confidence and ability in their three years at Edge Hill. **Steve Suggitt** is one of those people captured by Edge Hill; he has remained for twenty-eight years. Appointed from Liverpool University, Steve introduced foreign fieldwork, to Norway, and has initiated new degrees, in Earth and Environmental Sciences, Physical Geography and Geology, together with minor course in Geology. Remembering many fieldwork locations, he reflects on the joys of being outdoors on geographical research. **Nigel Richardson** current head of geography, now called Natural, Geographical and Applied Sciences (NGAS) since it combined with the science department, explains how raised peat bogs provide a record of past climate and of pollutant particles falling from the atmosphere, and have been a major focus of geographical research. Returning to a familiar Edge Hill theme of education and historical geography, **Charles Rawding** appointed from a school in Lincolnshire to deliver the new PGCE in geography to train secondary school geography teachers, has now trained over 150, some of whom are themselves mentors in host schools. In between dashing, in a very dashing yellow car, to schools all over the North West, Charles teaches on third year modules, in historical geographies and tourism, continues his own research and co-ordinates the local Geographical Association. **Tasleem Shakur,** appointed to replace Paul Gamble, was born in Chittagong in Pakistan, the place in the world said to have the highest rainfall. By his observance of the Muslim calendar he educates and widens our horizons. In describing his encounters with geography through the textbooks of Stamp and Monkhouse and contacts with 'Englishness' in the colonial setting of the former East Pakistan, he discusses a heady mix of learning, appearing in Shakespeare productions and listening to Western pop music, before coming to Britain to complete his education in architecture and planning. He offers an insight into his 'transformation' into international development geography. He has strong overseas links and has supervised many PhD students. Tas's laughter echoing round the corridors of the geography building keeps us all sane in the face of increasing bureaucracy. **June Ennis** recounts how her work as departmental administrator since the mid 1990s has rubbed off into environmental recycling activities at home, knowledge of the processes operating on the coast, friendship with staff and students, organising fieldwork, managing finances, but above all acting as social secretary.

Geography United

Changing technology has transformed the job of **Ann Chapman,** geography cartographer. Hand drawing, drawing boards, Rotring pens and Letraset have been replaced by computer programmes, designing Web pages, producing the departmental newsletter, plus some teaching on the use of the OS survey digitised map system, Digimap, while maintaining, cataloguing and supervising loans of the department map collection. Intending to stay for only a very short time, Ann reflects on just what it is about 'the very peculiar place' that has held her in the geography department for all this time. **Peter Stein,** GIS support technician for geography, recounts how far ahead of the field Edge Hill was and has remained in its provision for students. **Gerry Lucas** appointed in 1992 as GIS specialist explains the rise of the 'geo-information' branch of geography, enabled by improved computer software. He is also a geologist, whose current interests in geodiversity and geotourism make him a much in-demand speaker on cruise ships. Gerry encourages students to manipulate satellite images, and leads them in learning computer-mapping packages, in digitised photography and poster presentations. His assessment methods are innovative, causing students to spend many hours in the geography computer room, though he has described some of them as 'lounging there in Graeco-Roman style' (quote from a former student). **James Curwen** reflects on the challenging experiences and interesting places his work as a geography technician have involved; with research students drilling for peat samples in lakes, driving to the Outer Hebrides, and organising a group of Chinese planners across the country.

Sylvia Woodhead appointed as a human geographer at Edge Hill, despite her previous experience and research in physical geography, comments how the late 1990s 'cultural turn' in geography, a celebration of the interactions of people and the environment, seemed a return to the real essence of geography. She recounts her interest in designing innovative assessments and evaluations. Fieldwork in The Netherlands has led to admiration for the Dutch focus on protecting their natural environment, and in clear plans for sustainable living. She explains how sustainability provides a set of moral imperatives for 'treading lightly on the Earth', and how we should not steal from the future. She describes her latest research in developing indicators of sustainability that anyone may use to assess what is or is not sustainable in their lifestyle and surroundings. **Ann Worsley** came to geography through explorations, imaginary from an early age, which engendered a love of maps and 'foreign parts', via natural history type family holidays and school in Liverpool, inspired by her geography teacher. How often this theme is told. On meeting Professor Frank Oldfield she learned how to identify pollen grains, and went to Papua New Guinea, meeting tribes and investigating their relationships with their physical environment. She stayed in a hut and was made an honorary warrior. Ann returned to research later in the 1990s, teaching pollen analysis, before starting at Edge Hill, part time at first. Ann is a physical geographer convinced of the unity of geography, she leads many modules; physical geography methods, biogeography and Coastal Zone Management. For the latter she introduced an innovative mini-conference, with a professional audience, as part of the assessment. It is a frightening but well worthwhile experience, one to which students rise admirably. Ann also organises many coastal conferences, is a member of Sefton Coast Partnership, and has two new PhD students, Vanessa Holden and Ann Power. From her desire to be an explorer, she now delivers modules via e learning by WebCT, though she concludes that geography is exploration.

Vanessa Holden describes her research on the 'hinge line' of Britain, the muddy foreshore and salt marsh of the south Ribble estuary. Vanessa started her research with a desk study using GIS and aerial photographs, then fieldwork for ground verification. Her aim is to establish a database of rates of sedimentation by looking at changes along the coast on air photos, historic maps and charts and by using various artificial marker horizons and environmental magnetics testing. The aim is to produce a permanent record against which future change can be assessed and modelled. **Ann Power** describes how her research, funded by Halton Primary Health Care Trust, uses sediments in lakes as a record of air pollution, to link with human health. This research is innovative in using health records to make links with disease and mortality trends. In investigating the complex relationship between humans and the environment it is truly geographical. These personal perspectives of geography illustrate that the best definition of geography is that 'geography is what geographers do'. The story of geography at Edge Hill presents a number of interweaving threads: the Edge Hill- Liverpool link, the physical-human geography relationships and the making of geographers.

Geographers at Edge Hill believe they are delivering modern, vibrant and interesting courses in geography, which give students that 'edge' as they go out into the world. As recounted there is a long background of geography teaching at Edge Hill, where many people have made contributions, past and present, and the place has engendered some strong feelings. Certainly the College stimulates strong feelings among its students. Great loyalty and affection are displayed by students of all years-the best years of their lives.

As time passes those who experienced Edge Hill in the past are inevitably growing older, and their numbers are declining. Their memories offer precious links with Edge Hill's past, but have in some cases proved difficult to locate. Like all research, the investigations for this book have proceeded along many blind alleys, searches have proved barren, or inconclusive, or contradicted previous findings. More questions have been raised than have been answered. It is still not certain when and why the wartime Block A, now housing geography, was built. Supervised by Lieutenant Hawthorne of Burscough Barracks in 1940, it was said to be accommodation for nurses, though evidence shows that the Queen Alexandra's nurses lived in Stanley Hall in the main College building. Whether the geography ghost is a British or Polish soldier remains uncertain. Various people report hearing coughing late at night, or doors that mysteriously bang. The story that follows is comprehensive but not complete. It is hoped that readers will find it intriguing, interesting and inspiring.

Prologue

'Are you Human or Physical?'

Vanessa Holden

'Have you thought of Edge Hill?' asked the Careers Advisor. I gave a blank look and shook my head. 'Edge Hill in Ormskirk?' I was still none the wiser.

That was the first time I'd heard of Edge Hill, during my Careers Guidance interview in the autumn of 1991 during my A-level studies in the Midlands. I explained that I wanted to study physical geography with biology but that not many institutions seemed to offer it. The Advisor explained where Ormskirk was and showed me the Edge Hill prospectus with the course outlines. It sounded ideal, and sure enough, after a visit to the open day, Edge Hill had moved to the top of my list for places to study. It was the campus that first struck me, with the entrance to the main building being pretty impressive compared to so many other universities I'd visited.

'But Edge Hill isn't a city centre campus.' commented my friends who were applying to Manchester, Birmingham, and other 'big' universities. But that was exactly its appeal for me - the fact that it wasn't in the middle of a city centre. Certainly, if you wanted a city centre, it was one short train journey, but for me, I didn't want to be studying the natural, physical environment surrounded by tall buildings, concrete and tarmac. I wanted green space. The whole point of me studying physical geography was to appreciate the physical environment - so for me, Edge Hill was ideal.

Being a very quiet and shy eighteen year old, I was extremely apprehensive about moving away from home and away from friends and family to start three years of study. However, the friendly and supportive atmosphere within the department meant that I soon settled in, and thoroughly enjoyed my studies from the start. The size of the department compared to many accounts from friends at larger universities gave a very close-knit department, with the staff being very approachable. I still distinctly remember the first few days of sorting out modules and timetables and the like, and getting to know other fellow first year geographers, with the usual ice-breaker question being 'Are you Human or Physical?' My preference was definitely Physical Geography. I especially love the coast. Any opportunities to follow my own research during my studies were centred on the coast - wave and beach formation, sand dunes, sea defences, and now salt marshes. With Edge Hill being only a few miles from an incredibly varied stretch of coastline, it's ideally located for my research.

The course I studied - Geography (major) with Biological Science (minor) - gave me a real chance to study the subjects that really interested me. The different styles of the lecturers of course made it all the more interesting. Especially the lecturer who regularly showed slides (this was before the days of PowerPoint) throughout the lecture, so insisted on complete darkness in the room... great slides, but it was interesting to try and take notes in complete darkness. It was always useful to look back at your notes in daylight and to see if you'd actually managed to write on the paper. The modules on offer covered a range of topics and always seemed to be well structured and taught, although my friends and I do still remember the Statistics module scheduled for 3pm until 6pm on a Friday afternoon. What a start to the weekend, but at least it was memorable.

The campus has developed a lot since I was an undergraduate. There have been a number of new buildings constructed, and the facilities on offer to students have increased tremendously. I somehow managed to avoid the sports facilities for the three years I was originally here (with the exception of the gym for exams) but the new facilities really do encourage you to at least think about keeping fitter. The expansion of the library and IT facilities are also very impressive, with nearly all the journals now being available electronically and remotely. When I think back to my undergraduate days, when email was a bit of a novelty and not many people knew what the Internet was about. I do have memories of a friend discovering email and 'talking' to another student who was doing a similar course to him, and seemed to have a lot in common, which made sense when they realised that they were actually both at Edge Hill and emailing each other from the same library.

Some of the more memorable moments of the degree were of course the Field Trips. Trips to the Lake District involved staying in Ambleside, with a walk across town every morning for breakfast. There were plenty of complaints about this at first, but we soon started to look forward to our morning walk with food at the end of it, especially as it was downhill. One particular excursion that really sticks in the mind was a small group of us being dropped off in the middle of the Lake District, in theory to follow the course of a river through its catchment. However, when dropped off with one very badly copied map, we were told to meet up at the local pub by 2pm. Sounded reasonable until we asked where we were starting off from on the just about discernable map, to which the last words from the (nameless) lecturer in charge were … 'oh, you're not on the map, you're just off the edge somewhere. See you at 2pm.' That was an eventful day. Probably the most notable field trip for me was to Mallorca during my second year (Fig A). It was the first time I'd flown, but somehow my friends managed to keep me relatively calm during the journey, although thinking back now I'm pretty sure I must have been a complete pain for a few weeks before. It was an early morning flight, and as we sat on the plane waiting for take off, an announcement came over the tannoy: 'would Vanessa Jones please raise her hand'… the plane went quiet, and the expression on my face must have been complete horror. I (slowly) raised my hand, and a hostess came over to me. What a relief when they just wanted to check where I was sitting, as I had an awkward diet and they'd organised me a special breakfast. Marvellous. On arrival, the weather was gorgeously hot and sunny all week, but guess who forgot their sun block. Being of a 'delicate' complexion, sunlight and I don't really mix, so I scrounged sun cream from friends all week. This was only a problem when I borrowed one particular friend's sun cream, only to later discover it was Factor 2, and was about as effective as water at preventing me from getting sunburn. There were also some interesting experiences with beach surveying that I won't go into in any more detail, but suffice to say involved very tall trees and very small sand dunes. During the trip it was compulsory to

Fig A On fieldwork in Mallorca in 1994

Source: Vanessa Holden

complain about the lecturers' driving - although considering some of the roads and some of the minibuses, they actually did very well. Again, we soon all got into the routine of the trip, stopping off every morning at the local supermarket to stock up on lunch supplies. It was during one of these stop-offs that we found the white wine for equivalent of about 55p a bottle, where we decided that it was almost as cheap as the bottled water, and where my friend reckoned that if it tasted really bad we could just wash our hair in it... it wasn't at all bad though. By the end of three years, I had lost count of the number of times that I had got soaked, windswept, muddy and fallen asleep on minibuses, but that's what field trips are all about.

From Edge Hill to Liverpool and back

Jamie Halsall

'Geography is concerned with the description and explanation of the areal differentiation of the earth's surface.' (Harvey, 1969, p3)

There is no doubt that geography is an 'encyclopaedic knowledge' of 'spaces' and more importantly 'places' (Holt-Jensen, 1981): both of these are key words in the subject. Everyone at some point has studied geography whether it is historical landscape, population change or environmental disaster in a particular place. Geography is an amazing subject to study, yet is regarded by many people as a traditional subject. Unwin (1992) has described geography as 'one of the oldest forms of intellectual enquiry'. This article is about how I became interested in geography and how I feel privileged to have studied at Edge Hill.

Before I came to Edge Hill my main aim was to study Information Systems. At school I was not particularly good at geography, and I only managed to obtain a grade 'C' in GCSE. Throughout my A level Geography studies I enjoyed learning the theories behind population, coastal processes and the management of cities.

In August 2000 I was offered a place at Edge Hill to study for a BSc in Information Systems, and I decided to take Management and Human Geography as my minor subjects. I remember going to speak to Sylvia Woodhead about the course. I can still remember the conversation today, she explained why I should study Human Geography as a minor subject and strongly advised me to purchase a book, *Introducing Human Geographies* by Cloke et al (1999) which I duly did. The illustrations in this book, I remember one of 'Swampy' in a hole during an environmental protest, really inspired me, and the cartoons made me look with fresh eyes at cartoons in newspapers.

I attended my first geography lecture in mid September 2000 and remember sitting next to Jon Easthope. Tasleem Shakur sat next to Jon and me and started to talk about how he had become addicted to the computer game *Lemmings*. At the time I thought that Tasleem was also a new Human Geography student. A week later I discovered that he was in fact Dr Tasleem Shakur, a senior lecturer and Director of the International Centre for Development and Environmental Studies. I was struck by his modesty and his accessibility, this and his sincerity and approachability continue to the present day (Fig B).

Fig B Dr Tas Shakur in Amsterdam's cultural quarter

Source: Jamie Halsall

By April 2001 I had decided that I was not enjoying my studies in Information Systems and asked if I could transfer to geography and make it my Major subject. I am indebted to Dr Nigel Richardson who persuaded Academic Registry to transfer me to geography. I did not realise at the time but this was probably the most important decision I have ever made. The reason I transferred to human geography was that geography is relevant to everyone. The parts of geography that interest me the most are the issues, the social spatial relationships with people, surrounding 'Cyberspace' and 'Virtual Geographies'. The theoretical debates in human geography fascinate me, such as Time-Space Compression and the Shrinking World. These ideas encompass the effects of modern transport and computer technology in making the world apparently smaller. An email to friends in the States takes four or five seconds whereas a letter used to take as many days. In April 2003 I completed a dissertation titled 'Social Spatial Behaviour in Communication Technologies', which was supervised by Dr Paul Rodaway. This study used a focus group of people invited to my house – I gave them a cup of tea and a Rich Tea biscuit. I asked questions about their use of technology such as e-mail. I then analysed the tape-recorded data. Despite the middle-aged nature of my sample, I found them to be very reliant on and almost addicted to modern technology.

Fig C A glimpse into the Third World: recycling in a bidonville outside the walls of Marrakech, Morocco.

Source: Jamie Halsall

In my course I attended two geography field trips, to the Lake District and Amsterdam. When people ask where were you on September 11th 2001? I answer: I was on a salt marsh in Grange-over-Sands. I can remember a fellow student receiving a text message that two planes had crashed into the World Trade Centre in New York. Back at Castle Head Field Centre we met up with students from other universities and that night, as we gathered around the television watching the news bulletins, we debated the implications that the day's events would have in the future of the world in terms of economic and political geography. The Amsterdam field visit was also enlightening in that it showed how human geography actually works; I found it exciting and exhilarating. For my Amsterdam research project Jonathan Easthope, James Caldwell and I examined housing differences either side of Vondel Park. The park did indeed appear to act as a barrier between a wealthier southern district and a more working class northern area of the city centre.

In October 2003 I had the opportunity to visit Morocco, my first visit to a non-European country. This field trip gave me a better understanding of the issues of the Muslim sub-culture. Before the visit I had expected to encounter more poverty in rural areas. However the wonderfully located *Kasbah du Toubkal* in *Imlil* village in the Atlas Mountains, a winner of the Green Globe for Sustainable Tourism, is an enlightened Berber settlement, and the village even had an Internet café. I was more shocked at a visit to a shanty town outside the walls of Marrakech. We visited a market where meat for sale lay on the ground covered by flies, while rubbish ready for recycling lay piled around (Fig C).

Following my graduation from Edge Hill, I decided to study at the University of Liverpool for an MA in Geographies of Globalisation and Development, because of my research interests in cyberspace within the context of globalisation. What is surprising is that both institutions have a reflexive relationship in that most geography lecturers who currently teach at Edge Hill gained their qualifications at the University of Liverpool. For example Steve Suggitt, Gerry Lucas and Sharon Gedye did their undergraduate degrees there, while Nigel Richardson, Sharon Gedye and Ann Worsley undertook their PhD research in Liverpool's geography department, whilst Dr Tasleem Shakur, mentioned earlier, completed his PhD research in the department of Civic Design. So when I progressed to Liverpool I felt that I was keeping up the tradition.

When I studied for my MA course I decide to go deeper into cyberspace. My first essay was concerned with how cyberspace has transformed the trade union network within the political and economic world. My second essay argued that cyberspace fundamentally has transformed the spaces of work, especially when it comes to teleworking and telecottaging. For my dissertation I decided to research the relationships between cyberspace, identity and globalisation. This research helped me understand cyberspace in a more conceptual way, for example, how it interacts with the economic, social and cultural processes of globalisation and how cyberspace affects human behaviour in the modern world.

I still have strong links with Edge Hill because for the last two years I have been contributing to teaching undergraduates human geography, for example in the method of focus groups which has formed part of my BA and MA work. I would like to take the opportunity to thank all the members of staff who taught me geography at Edge Hill. They were great days, which I will never forget and provided a firm foundation for a lifetime's interest.

References

Cloke, P. Crang, P. and Goodwin, M. (1999) *Introducing Human Geographies,* London, Arnold.

Harvey, D. (1969) *Explanation in Geography,* London, Arnold.

Kitchin, R. (1998) *Cyberspace: The World in Wires,* Chichester, John Wiley and Sons.

Holt, A. J. (1981) *Geography Its History & Concepts,* London, Harper & Row.

Unwin, T. (1992) *The Place of Geography,* London, Prentice Hall.

Part I Beginnings

1: Empire Days 1885-1933

This chapter opens with Edge Hill-Liverpool links. It reviews the founding of Edge Hill Training College at Durning Road in Liverpool as a college for women, and explores the connections with the philanthropist George Holt. Using the annual Guild of Students

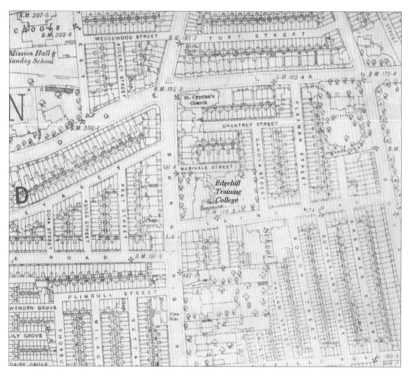

Fig 1.1 Edge Hill Training College at the junction of Durning Road and Clint Road, Liverpool: Reproduced from the first edition 1893 OS Lancashire sheet CVI.15 1:2500

Source: NGAS Map Collection

magazine, plus Fiona Montgomery's history of Edge Hill from 1895 to 1997, to which readers who wish to know more about these early years are recommended, it has proved possible to discover some elements of the geography teaching, the teachers, their lectures and details of fieldwork visits. It is significant to note that Edge Hill was providing geographical education for women before they were accepted into the Royal Geographical Society.

The George Holt connection

Predating Lancaster University by eighty years, for the first forty years after its foundation in 1885, Edge Hill College was based in the Edge Hill district of Liverpool, at 78-79 Durning Road (Fig 1.1) in the former private house of George Holt senior (1790-1861) who, after moving from Rochdale to Liverpool, made his fortune as a cotton broker. In 1820 he secured his position by 'an advantageous marriage' to Emma Durning, who came from a rich and long established Liverpool family, after whom the road was named. Both were active in local politics, education and philanthropic schemes.

Fig 1.2 George Holt timing an experimental physics machine he had bought for Liverpool University

Source: Liverpool University Special Collection

George Holt was an important man in Liverpool, an entrepreneur, with family links to the Blue Funnel shipping line, and a philanthropist supporting local education. He was responsible for the construction of the first India Buildings in Water Street, Liverpool, where the Blue Funnel line had its headquarters. His ships had names from Homer's Odyssey. **The Blue Funnel Line** was the world's first regular long distance cargo liner service. It played a major part in Liverpool's shipping activities, and led to the development of the Chinese community in Liverpool. George Holt II became a partner in

1845 in founding a shipping line (Lamport and Holt) that traded with India, South Africa and South America, and was a pioneer in development of steamship services to Brazil. His younger brothers Alfred and Philip Holt became important shipping merchants with the Blue Funnel line. The Holt family were among the first in Liverpool to employ women in their offices. George Holt II, born in 1825, married Elizabeth Bright: their only daughter Emma was born in 1862. Like her father Emma Holt was a noted philanthropist and supporter of Liverpool University.

Fig 1.3 Edge Hill College's first home in George Holt's former mansion, Durning Road, Liverpool in 1903

Source: Edge Hill archives

George Holt, one of six trustees of Liverpool University who gave £4000 in 1884 (Kelly 1981) was described as 'one of Liverpool's merchants, whose views on art and culture altered the character of the town' (Hyde 1971:43). A full account of his philanthropy to Liverpool University is found in a speech by Lord Derby on the late Mr George Holt, of 'no previous academic associations, a constant and steady supporter', in the 1896 edition of the University of Liverpool Students magazine, *The Sphinx* (vol. III No27: 164).[1] His numerous gifts included a Tutorial Scholarship Fund £2000, £1000 to the library, £1000 to the Victoria building, plus £1000 for 'perfecting the vestibule and staircase'. He founded a Chair in Physiology, and gave £10,000 with a further £5000 for lab 'adjuncts (Fig 1.2). He founded a chair in Pathology with an endowment of £16,000. He furnished, equipped and presented the Ashton Hall Bacteriological Lab, probably with £5000, and bought adjoining houses on Brownlow Hill. He offered £1700 to the building of the Medical and Anatomical Buildings, and gave £1000 for the Natural History Museum. His gifts to the University totalled over £40,000. He also paid the University College's debts.

The Edge Hill-George Holt connection is intriguing. The College was established in his former house on Durning Road, but much more is known about Sudley House, Aigburth, now maintained as a museum, where he moved in 1884, when he and his wife both 'semi-retired' to concentrate more on their art collection. At Sudley House George Holt organised many radical alterations: he added bay windows, a tower, extended the house and added wallpaper, panelling and marble fireplaces probably from Italy. On her death in 1944, their daughter Emma Holt bequeathed Sudley House and its painting collection to the City of Liverpool, which now maintains it as a museum. It has not been possible to discover if George Holt's philanthropy extended to Edge Hill, if he donated his former house, or supported Edge Hill in other ways. He is listed as one of Edge Hill's trustees who donated £5 each.

As befits George Holt's wealth, the Durning Road **building** was very fine, in a classical Georgian style with large symmetrical windows, a balustraded roof and a columned main entrance, with a separate front entrance for students (Fig 1.3). There was a surrounding garden, a gymnasium and tennis courts. The house was adapted for College use by the addition of a new wing with classrooms and dormitories, and a second new wing with a common room and better dormitories was opened in 1891. In 1895 another small building was added, and the kitchen extended. 1896 was 'a memorable year', with a further extension, electric lights, and telephone[2]. A new wing was added in 1904[3,] and extra room acquired by renting two adjacent houses. Despite these frequent extensions there was still insufficient room, and some students were boarded out.

---

[1] Liverpool University Library Special Collection
[2] The Story of the Evolution of Edge Hill College
[3] Edge Hill College Magazine, No 13, 1904

A College for Women

Edge Hill was originally a non-denominational educational College for **women,** initially forty students, all residential, though probably from Liverpool and the north-west. Montgomery (1997:2) notes that students were rigorously selected from a pass list of the Scholarship Examination. They had to have good health, as teaching was regarded as a strenuous career for which (female) stamina was required. Certainly hygiene is repeatedly mentioned in College newsletters. Students, all female and from dominantly middle class backgrounds, were issued with a comprehensive list of clothes to bring, 'to ensure uniformity'; a sailor hat with the College band and badge was compulsory weekday wear. Pictures show serious faced young women in high necked, long sleeved light coloured blouses with floor length dark skirts. It was in this outfit that they went on an annual College Picnic (Fig 1.4). Travel would have been by train and ferry[4], and finally walking; in 1906 the annual picnic went to Thurstaston, also on the Wirral. Students were accustomed to making their own entertainment; the Edge Hill College Magazines, started in 1892, say much about the plays staged most weekends, and also record the Saturday evening lectures given from 1891, on Literature, History, Travel, Science, Music and the Arts; Geography possibly represented by the travel category.

Early College life was very Spartan and inflexible. A rising bell sounded at 6.15am for prayers at 7am before breakfast at 8am. Students then made their beds, dusted and tidied their rooms before morning work from 9.00am to 12.00pm, then half an hour recreation before dinner. Further study from 2.30 to 4.30pm was followed by tea at 4.30pm. Students were made to go out for a walk from 5.00 to 6.00pm in the fresh air of the pavements of Edge Hill. After study from 6.00 to 8.00pm, supper was at 8.30pm, prayers at 9.00pm, silence at 9.45pm and lights out at 10.00pm. The age of majority at the time was twenty-one, which meant that students were regarded as minors and they were constantly supervised. Saturday afternoons were free, while Saturday evenings were spent in needlework or drawing while the Principal read aloud from 'an interesting or instructive book' (Montgomery 1997:28). Accommodation was very basic with little privacy. Students fetched their (cold) washing water, while baths were taken according to a timetable allowing each student fifteen minutes.

---

[4] They could have travelled by rail to the Wirral: the Mersey Rail Tunnel opened in 1903. The first road tunnel, Queensway, was opened in 1934

Miss Butterworth, an Edge Hill student of 1902 to 1905, writing in Edge Hill College Magazine 1955, remembers the 'stress' of Durning Road, 'the long treks on foot or by tram to schools, to city swimming baths, a very long journey to a very distant hockey field for a very short game of hockey'. Living conditions, society and technology were greatly different from now; there were no cars, computers or consumerism.

Fig 1.4 Edge Hill College Picnic to Leasowe, Wirral in 1904

Source: Edge Hill archives

Not only were all the students and resident governesses female, but so too were the first Principals, though visiting professors tended to be male. Edge Hill College in Liverpool had three female **Principals.** A student from the earliest days of Edge Hill, under Miss Yelf, had two vivid impressions, of a sense of urgency in the pursuit of higher education for women and a sense of the high ideals of service required. The first Principal, Miss Yelf, retired on health grounds after five years, to be replaced by Miss Sarah Jane Hale, who died in post and was succeeded by Miss Smith; female teachers were not married at that time. Miss Smith had the amazing experience of being Principal on all three College sites, in Liverpool, Ormskirk and Bingley. Miss Yelf and Miss Hale had education backgrounds, while Miss Smith was a Cambridge mathematics graduate (Montgomery 1997:13).

Professor Kelly, in writing a history of Liverpool University's early days, notes that after the 1870 Education Act, which considered the training of teachers important, the University held Saturday morning courses, and after 1885 it was mostly Edge Hill Training College students who enrolled (Kelly 1981:76). In 1893 the University agreed to recognise a certificate for a University Extension course as part qualification for the Queens Scholarship examination, and the first of such courses for primary teacher training in 1894 was delivered in collaboration with Edge Hill. (Kelly 1981:120)

From 1894 **Miss Hale** introduced these three-year university courses for suitably academic students though third year students felt slightly isolated from the Edge Hill Community, as did day girls (Montgomery 1997:6). Students were in awe of Miss Hale, with her outspoken Victorian views that stressed duty and service. During her time Edge Hill was a place of 'stern endeavour, hard work and unwearied perseverance[5]'. She had a personal maid to wait on her at table (Montgomery 1997:40). Margaret Bradley (née Upton, student 1949-1951) reflecting in 2002[6] recalls that her grandmother was 'in service' at Edge Hill College in 1885, and became maid to the Principal. When she left to marry (a woman getting married at that time gave up work) she was presented with a Wedgwood tea service which the family still own. Her mother Margaret Lipton (née Baker) also went to Edge Hill from 1914 to 1916 and gained a Certificate of Education. This kind of family loyalty to the College is typical of Edge Hill students.

Miss Hale founded the **Guild** of former Edge Hill students, whose annual newsletter started in 1892 shortly after her appointment, and still continues today. The first cohort of forty students was 'sent out into the world' in 1886. Students were expected to take teaching posts away from Liverpool, hence the Guild for many years had branches in London, Manchester, Birmingham and Sheffield. The Guild later raised over £1000 to furnish the Hale Memorial Hall at the new Lancashire Education Committee building in Ormskirk, to preserve the name and memory of the Principal who had done so much for the early days of the College.

Geography was one of the 'professional' **subjects** included in the teacher training syllabus, from the early beginnings of the College. Before 1904 the Board for Education formulated training college syllabuses, and initially the College syllabus followed the same lines as

---

[5] Edge Hill College Magazine, No 19 1910
[6] Edge Hill College Guild of Students Newsletter 2002

the Scholarship Examination, though in more depth. Students studied practical teaching, reading and recitation, arithmetic, music, grammar, literature, geography and history. From 1907 the Scholarship Examination was replaced by the Preliminary Examination for the Certificate, when geography was taken in the second year. The large number of subjects studied was reduced in 1913 when geography was classed as 'General Studies, where students had to take three or two of English, History and Geography, Maths and Elementary Science' (Montgomery 1997:15). All indications were that the College was well ahead of its time in providing a high level of education to women. In 1910 the College had sixty-eight girls, thirty-eight on a two year course, seventeen day students and thirteen University students with a three year course, while in 1917 Liverpool and London Universities created the first full University chairs in Geography.

Edge Hill Principals (Geographers indicated*)

Durning Road, Liverpool

Miss Sarah Jane Yelf	1885-1890
Miss Sarah Jane Hale	1890-1920

Durning Road, to Ormskirk & Bingley

Miss EM Smith	1920-1941

Bingley to Ormskirk

Miss Elsie Butterworth*	1941-1948
Miss Margaret Bain	1948-1964
Mr Ken Millins	1964-1979
Miss Marjorie Stanton	1979-1982
Mr Harry Webster	1982-1989
Professor Ruth Gee	1989-1993
Dr John Cater*	1993 - present

Imperial Dogma

An idea of the likely geography syllabus at Edge Hill from 1885 may be gained from L W Lyde writing *A Syllabus of Geography* in 1898. He considered geography as good training for the mind, and believed that a school syllabus should start from what pupils ('boys') ate, wore and heard discussed. He was a strong advocate of using an atlas to 'absorb maps insensibly... every boy has an atlas of his own in which he looks up every country mentioned'. Geographical education was part of a broader culture, imbued with the all-pervasive central ideology of imperialism and support for the British Empire (Ploszajska 1999).

Edge Hill College Magazines express the popular support for Britain's acquisition of territories overseas; beliefs and attitudes focussed on patriotism, loyalty to the monarch and belief in the destiny of Anglo-Saxons to rule the world. Geography conveyed knowledge of the world and was seen as useful to the British Empire; it was designed to assist exploration; hence much emphasis was put on mapping and description of countries. The Victorian belief in British racial supremacy was also linked to environmental determinism, a view that humans are limited in their endeavours by their environment, thus Africa 'the dark continent' was seen to be lacking in development because of its limiting hot climate and impenetrable tropical rain forests. Britain's climate by contrast was regarded as 'neither too hot nor cold', while, in this type of climatic determinism, New Zealand was considered the 'healthiest' part of the empire.

The **Royal Geographical Society** (RGS) had been founded in 1830 to publish the results of (male) explorers, but only in 1870 had become interested in education (Driver 1995:404) though at first its focus was on white elite male public schools. Geography education, for women, at Edge Hill may well have been a great contrast to the situation in British public schools, where a report by an Inspector of Geographical Education, appointed by the RGS, found a poor situation, where the privileged male minority received an education which 'rarely included a systematic geography syllabus, maps, visual aids and textbooks were generally scarce, dull and obsolete' (Ploszajska 1999:4). 'Traditional Geography' from 1885 to 1915 consisted of rote learning of capes and bays. Ploszajska (1999:9) describes regimented rows of pupils, the teacher maintaining a stern watch over catechismal learning. Key facts, league tables of principal towns, imports and exports, were listed and memorised. Interest was held by fear. Geography came to be regarded as an effort of memory (Marsden 1976).

Halford Mackinder, proposed for the RGS in 1886, was influential in promoting more progressive methods in UK geography, and he became part of the movement for educational reform, and championed the use of visual imagery in geography lessons. He was convinced that geography was one subject, and believed in teaching geography as a whole. 'There is no geographical region less than or greater than the whole of the earth's surface', Mackinder 1910, quoted in Gilbert 1960. In 1893 he became involved in the founding of the **Geographical Association** (GA) to stress the importance of teaching geography in schools, though he believed the 'geographical gaze' to be the specific privilege of white European males (Ploszajska 1999:161).

An ability to see a vivid image of distant and unknown places in the mind's eye was considered to be among the most important of a geographer's skills, and one of great value to all citizens of the vast and diverse British Empire, Halford Mackinder quoted in Ploszajska 1999:83. The GA journal from 1901 disseminated ideas on map work and fieldwork and provided a lantern slide and book lending scheme. In 1902 Mackinder published his famous book *Britain and the British Isles*, a **regional geography** which included climate and geomorphology, and which contained the Tees-Exe division of Britain into Highland and Lowland. Mackinder was interested in geopolitics and his 'Heartland' theory is still considered significant today. He believed that observation, cartography and teaching were essential characteristics of geography. Mackinder was said, by H J Fleure, to be handsome with flashing eye and a gift of oratory (Balchin 1993:3). Other influential regional geographers such as H J Fleure and Vidal de la Blache also regarded the natural and cultural worlds inseparable. Regionalism had been a dominant approach in Geography since 1890. E W Gilbert, giving the A J Herbertson Memorial lecture to the Institute of British Geographers (IBG) set up in 1930, comments on the powerful influence that the idea of a natural region had on the teaching of geography in schools and universities in Great Britain. Herbertson deplored nineteenth century tendencies to divide geography, particularly into political units, and instead devised the concept of natural regions (Marsden 1976). Geography was considered to be 'the art of recognising, describing and interpreting the personalities of regions'; which were said to be identified by names, like the Lake District or North West England. Physical geography was considered essential, and general world geography with a special study of a region was the ideal to aim for.

Women and the Geographical Association

The question about whether the GA admitted women teachers from the beginning has been a surprisingly difficult one to answer. The GA was certainly founded by men, albeit including some like Douglas Freshfield who were very pro the admission of women to the RGS, and it remained for a long time male dominated. Balchin 1993 describing the setting up the GA includes mostly photos of men, but doesn't actually say if women were admitted from the start. By 1905 only a third of GA members (153 out of 486) were 'ladies', and it was not until 1963 that a woman, Professor Alice Garnett, was President of the GA, and she was the only female President out of fifty mentioned by Balchin.

The GA was founded in 1893 by a meeting of Public School Masters (eleven men, Balchin 1993:3) the related correspondence (thanks to Frances Soar) makes frequent mention of 'masters' and what 'boys' should study at school, but there is no mention of any 'sex' restrictions on membership. On the contrary several women were listed as attending the first AGM in 1894, and the 1897 Annual Report lists several girls' schools in which the GA was 'represented'. In the 1900 AGM a motion was carried 'that the Association shall be open to all Teachers of Geography', in effect to allow primary school teachers to become members, though this is tantamount to admitting women, as primary teachers were dominantly female.

The origins of the GA are strongly linked to education: it was set up because the RGS didn't take teaching geography or promoting appreciation of geography for all very seriously. Some early members of the RGS ridiculed attempts to bring geography to a wider audience, by the use of modern technology of lantern slides, calling them 'Sunday school methods'. The GA, whose mission now is 'to further the teaching of geography and the value of learning geography for all', has a focus of practical teaching and learning.[7]

Learning by Reading

Edge Hill students would have been familiar with the type of geography **textbook** (still the most widely used resource for teaching in British schools, despite much criticism, Marsden 2001) presenting concise, unquestionable facts, all to be mechanically committed to memory, many of which were slim, cheap and designed for homework exercises. Geography textbooks, especially those designed for elementary schools, were very much influenced by the political climate of the time (Graves 1996). Some texts expressly prepared pupils for examinations, giving odd assortments of facts, and information on relief, climate and rivers, while one 1881 text divided 'productions' into Animal, Vegetable and Mineral, and included sample examination questions, a couple of which are shown below.

> 5. How is India divided for administrative purposes? Name the chief races of India.

> 6. What are the chief lines of railway in India? Name the principle stations.

---

[7] Telephone interview with David Lambert 18 January 2005

Herbertson's ideas on natural regions greatly influenced later geography textbook writers, Newbiggin, Unstead, Stembridge, Stamp and Pickles. However some of these texts presented oversimplified, static and deterministic generalisations and were not considered much of an improvement over the older capes and bays geographical tradition. Walford 1995, when analysing a sample of popular frequently reprinted geography textbooks of the interwar period, found that typical texts such as Fairgreave and Young 1928, had eighty-seven per cent written text with only the occasional sketch or photo, partly limited by costly letterpress production methods of the time. Geography was then seen as the accumulation of knowledge and regional geography held centre stage of the school curriculum.

Syllabus of Geography Lessons given in a Lancashire village in 1905

1. *Plan of schoolroom*: position of windows, door, fireplace and furniture

2. *Plan of school garden*, including children's plots

3. *Plan of village*: main roads and landmarks

4. *Outskirts of village*: roads and railway cuttings

5. *Bird's-eye View from Church Tower*: to include direction and wind force

6. School Excursions: neighbourhood walks

7. *The Seaside*: drawing maps of the coast from Fleetwood, past Blackpool to Southport

8. *County of Lancashire*: using Philip's Wall Map and OS maps, map reading and moulding of the county in plasticine

9. *Northern England*: Lake District, Pennine barrier, 'other districts' for coal and shipping

10. Journeys to more distant parts of England: route to London

11. *England*: position and size on a globe and map of Europe

12. *Outside England*: mountains and rivers, 'children of other lands', their ways, dress and food - **hence** *(My italics - SW)* varieties of climate and productions, exchange and commerce.

In fact many schools used Geography readers of a more travel excursion type, which had woodblock illustrations, photos, maps and diagrams, and were designed to excite children's imaginations. So much use was made of fairy tales and adventure stories that in 1905 the Board of Education insisted that all Geography teaching must be firmly grounded in reality (Ploszajska 1999:102). In 1905 *The Journal of Education* published a syllabus, for eight to sixteen year olds, sent in by a former Head Mistress of a rural high school in Lancashire, which may be similar to the experience of Edge Hill students around the turn of the century. The syllabus lists twelve items, the first being a plan of the classroom, and only the last being outside England.

The syllabus took a concentric approach, starting with features familiar to the students, using the classroom to plot the course of sunlight over the day, with great emphasis placed on cartography and the production and understanding of maps. It was very local, based on personal observations and walks in the local area, the view from the church tower and journeys to school. There was much focus on Lancashire, with lesser treatment of the rest of northern England. The overview of London and the rest of England were very brief and only related to location, with minimal attention to the rest of the world. The last item on the syllabus shows an unquestioning acceptance of climatic determinism, from the suggestion of being able to deduce climate from clothes worn. Surprisingly colonies are not mentioned at all.

Attention to both physical and human geography is seen in the annual Edge Hill College Magazines. These give a fascinating glimpse of geography teaching at the zenith of the British Empire, which in the 1920s covered possessions in all five continents, encompassing one-fifth of the globe and 400 million people (Ploszajska 1999:23). Classrooms of the day would have featured a large wall map of the British Empire on the Mercator projection. Using the Annual Codes of the Board of Education Ploszajska has traced the development of school geography; in 1875 the New Code added the requirement of special knowledge of the County in which the school was situated, in addition to a geographical study of England. In 1885 HMI advised attention to English colonies. Amendments to the Education Codes in 1889 and 1890 encouraged the geography syllabus to emphasise connections between pupils' everyday lives and wider world affairs; a breakfast meal was often used to describe reliance on world commodities (Ploszajska 1999:53). By 1905 the Board of Education considered geography as the subject to bring school pupils into contact with the outside world, while plotting military wartime manoeuvres was extolled as making geography 'real'.

The 1906 Edge Hill College Magazine notes that a specimen lesson was given by Mr Cartlidge, Tutor for Geography and

Arithmetic, while in 1912 Mr Clarke of Clint Road School (Clint Road adjoined the Durning Road site) gave an account of the teaching of Geography in his school, and in 1913 Miss Winchester was mentioned as teaching Geography.

> He (Mr Clarke) brought many maps, models and other specimens of handwork done by his pupils. His address and illumination proved very helpful and showed the large amount of work on practical Geography which can be done by children. (Edge Hill College Magazine No. 21 1912)

Mr Clarke was obviously an innovative teacher, making use of models in the classroom. **Models** were part of a movement for active learning, helping make Geography lessons more enjoyable, appealing to children's tactile senses. In 1874 the Inspectorate had advised that schools should have faithful models of an actual glacier, volcanic island and mountain, though catalogue models were 'ideal' with greatly exaggerated (x8) vertical scales, causing much discussion over correctness of scale. Teachers were encouraged to construct accurate models from the Board of Education store, though by 1888 pupils' modelling in everyday materials such as sand and plasticine was a requirement of Geography lessons, said to train working class children for manual jobs (Ploszajska 1999:193).

Working in the Field

Although Liverpool was bombed by the Zeppelins, and 'we were compelled to shroud our lights by brown paper'[8], Montgomery (1997:37) noted that the **Great War** had little impact on Edge Hill students, though they helped the war effort by making garments for the War Service Bureau, and entertained soldiers from hospitals in Liverpool. In 1914 at Durning Road 'wireless telegraphy was added to the scientific apparatus, though for receiving rather than sending messages'. In 1915, the examination was changed to two grades of paper, Ordinary or Advanced levels, and, even though it was during the war, annual visits to the Walker Art gallery in Liverpool[9], and Geography excursions continued (Fig 1.5).

> The Advanced Geography Class under the guidance of Miss Deakin visited the submerged forest at Leasowe and also went to Roby to visit the junction of two streams. It was unfortunate that on the latter expedition one of the

---

[8] Edge Hill College Magazine, No 25, 1916
[9] Edge Hill College Magazine, No 24, 1915

streams was noticed to be flowing in the opposite direction, so that the right one had evidently escaped observation, but the interest and enjoyment of the ramble was in nowise impaired by the discovery (Edge Hill College Magazine No. 26 1917)

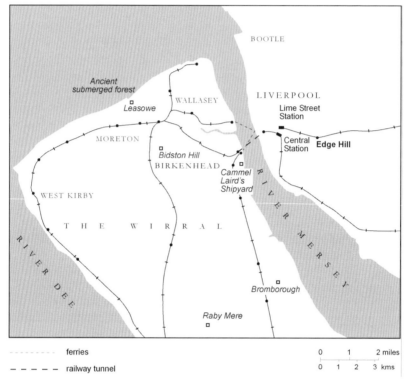

Fig 1.5 Geography fieldwork locations on the Wirral, 1917-1919

Map: Ann Chapman

Tracking the course of a stream was the type of adventure-inspired geographical fieldwork activity popular around 1910, though textbooks contained few female role models. (Marsden comments on the gender bias in geography textbooks, and quotes research by Bale 1981 that men outnumber women by 4:1 in the illustrations in geography texts he surveyed, and were often shown as subservient.) The importance of geographical knowledge in military manoeuvres was obvious, leading to increased public interest in geography during the First World War (Balchin 1993). Ironically it was during the Great War that the students had a geographical lecture; in 1916, Mr G E Thompson came to the College to talk about 'The Volcanic Eiffel and the Moselle', while the Principal, Miss Smith, showed slides of 'Reminiscences of Spain'. The wide foreign travels of Edge Hill Principals, as recorded in College Magazines, contrast strongly with the narrow geographical focus of rural school lessons. Several missionaries visited the College and gave 'interesting accounts of their work', while on the annual picnic to Eastham, students gathered flowers and grasses round Raby Mere (Fig 1.5). In 1917 students walked to Raby Mere from New Ferry instead of the annual picnic, while in 1918 Guild meetings continued, and some students volunteered to harvest flax (to produce jute sacking no doubt) in Bedfordshire and Somerset.

> The Advanced Geography Class under the guidance of Miss Deakin visited the submerged forest at Leasowe and had also been to Bromborough to see a remarkable fault in the rocks (Edge Hill College Magazine No. 27 1918)

Geography played a major role in peace treaty discussions, and delineations of boundaries after the war (Balchin 1993). On the declaration of peace an afternoon holiday was granted for Armistice Day in 1919 and Miss Hale's address spoke of better salaries and pensions for teachers, though conditions in the College worsened, with outbreaks of influenza. Edge Hill College had given up its hockey field in return for using Sefton Park, but was becoming more crowded, and plans were discussed for an extension in Clint Road. In 1919 the annual Advanced Geography excursion had taken place to Leasowe, with a train journey to Moreton and a walk to Leasowe, with only one casualty with a sprained ankle (no Health and Safety regulations then).

> A second Advanced Geography excursion, accompanied by Miss Deakin, took place in the Summer Term to Bidston Hill where evidence of ice erosion was found, and a fault was traced in the cutting in the Hoylake Road. Miss Deakin also showed lantern slides illustrating the work of ice and the economic Geography of South Lancashire. The slides were explained by Miss Deakin to the Geography classes of both years and it was much appreciated by all who were there. (Edge Hill College Magazine No. 28 1919)

Fieldwork is synonymous with geography, though in earlier days frequently described as excursions, as are those of Edge Hill which are typical of the urban to rural visit. Miss Deakin's trips seem to have had a physical geography and geology focus (Fig 1.6) there are initially no accounts of trips on the Mersey or visits to the docks. It is interesting that Edge Hill students went on fieldwork visits: In 1870

HMIs emphasised the value of school grounds for measuring dimensions and preparing plans, and by 1882, based on the local to global approach, it was compulsory for pupils to produce a map of the playground.

Fig 1.6 The Bromborough Fault: a roadside exposure of classic normal faulting, used as a fieldwork location since 1918

Source: Gerry Lucas

The use of **lantern slides** (early photographic slides, two inches square, glass mounted, and displayed by a projector) was advocated by the RGS from 1886, and was the focus of early GA activities, which produced sets, lent only to boys' public schools till 1900. Balchin (1990) argues that heated discussions around the use of lantern slides led to the origins of the GA. The very first lantern slide was a map of the world on a Mercator projection showing British possessions in red, and later a special photographer was commissioned to travel through the empire to make slide sets to which Mackinder wrote lectures. Lantern slides were regarded as the most useful educational aid in the inter war years, and were said to be a favourite means of learning geography (Ploszajska 1999: 169). The mix of geomorphology and economic geography in the lantern slides shown by Miss Deakin to Edge Hill students was typical of the time.

From 1925 BBC **Radio broadcasts** were 'radiated' to schools, and more schools listened to geography lessons than any other subject (Going 1938:261). Broadcast lessons were intended to 'supplement and not supplant the work of the teacher', but were felt to bring rural schools in particular 'into touch with the outside world', and to provide a 'connecting channel between the school and the mainstream of the nation's life'. It was felt that this 'novel experience' would interest children, though some early talks were very formal. In 1927 geographical Travel Talks for primary schools had supporting lecture notes and photos. In 1932 two Edge Hill students[10] had attended the BBC Summer School in Oxford to discuss the programme of talks for the coming season, though they reported on their return that few responses were made to their suggestions.

Moving pictures were supported by the Board of Education as early as 1914, though not intended to replace lantern slides, and from the early 1920s experiments with **films** 'brought the world' into geography classrooms. Various organisations; the Ministry of Farming, the Empire Marketing Board, the Imperial Institute, produced films lent free to schools, and in 1933, as Edge Hill was moving to Ormskirk, the GA recommended their use. Other organisations produced free adverts and posters, which were used on classroom walls. These pictured stylised imperial trade routes and production of raw materials for Britain, or were evocative images of cotton pickers in Sudan or the tea industry in Ceylon. While Edge Hill was in Liverpool, geography was dominated by ideologies of empire, environmental determinism and racial superiority.

Changes

Following the death of Miss Hale in April 1920 there were many appreciations of her life and work, claiming she was progressive and never afraid of innovation, a view disputed by Montgomery (1997:39). She was judged to have considered the training of teachers as her life's work. Under her guidance students were encouraged to take University degrees, while she made efforts for the enlargement of the College buildings, and fostered the Dramatic Club. It was resolved to name the new extension the Hale Wing after her. Students were very upset when she died, and one appreciation read:

> Without her, we should have missed many of the rare
> beauties of life which she led us all to know and appreciate.
> (Edge Hill College Magazine No. 29 1920)

---

[10] Edge Hill College Magazine, No 42, 1933

Miss Smith took over as principal, inheriting considerable financial problems, together with problems with the boilers and maintenance of the buildings. Her first report referred to 'a difficult year', and noted some changes. University courses became four years in length, because there was so much to fit in. Shortages of teachers possibly continued; staff were certainly required to teach a wide range of subjects. Miss Deakin, who had led geography field trips and who had been excused the additional duties of Music teaching in 1919, was now described as also teaching English, and gave an evening lecture on 'The Women of George Eliot'. Later entries in the Edge Hill College Magazines show her as a musician playing the violin. Gifts of books and magazines, such as the American Geographical Magazine, were gratefully received in the difficult post First World War years. On the anniversary of Armistice Day in 1922 many new societies started in the College. The Camping Society gave 'Scenes from Camp Life', there were fireworks on the Tennis Courts, and Miss Deakin played the violin for a rendition of *Hiawatha's Feast*. The Natural History Society met once a fortnight and arranged rambles to Hilbre Island, Knowsley, Raby Mere and Dibbins Dale (on the Wirral) while the Gardening Society planted seeds and bulbs and 'resuscitated' the lawns. The Cycling Society organised long rides to Freshfield, Delamere, Chester and West Kirby. Interestingly these new societies were not again referred to in College Magazines.

Edge Hill College buildings in Liverpool were old and poorly maintained; Inspectors made frequent complaints on their condition with demands for improvements to accommodation and cooking arrangements. In 1925 a Board of Education inspection completely condemned the buildings. The only solution was a new building in a new place. The College needed Local Authority help, and Edge Hill was acquired by Lancashire Education Committee, which agreed to provide a new building, and to continue the name and reputation of Edge Hill, as apparently requested by Miss Smith.

Edge Hill, Liverpool

Liverpool's Edge Hill district lies on a sandstone outcrop high above the city centre. In 1836 its railway station was built and in the early days all trains stopped at Edge Hill where a stationary engine hauled trains up and down the steep gradient to Lime Street station. Edge Hill station had huge marshalling yards, sorting trains from the docks via two tunnels. Edge Hill is also famous for the labyrinth of tunnels excavated from 1818 in the sandstone by teams of workmen employed by a retired tobacco merchant Joseph Williamson, nicknamed the 'Mole of Edge Hill'. Williamson's tunnels had no clear purpose, other than to provide employment during the difficult years after the Napoleonic Wars. Local enthusiasts have excavated rubble from inside some of the tunnels, and a visitor centre interprets his heritage.

Why was Miss Smith so determined to take the name Edge Hill: how distinctive is it? The answer is not very. There are many others both in England and abroad. Signifying the edge of a settlement, the name is frequently used for a hill or scarp, as in the gritstone edges of the Peak District. Perhaps the best known 'other' is Edge Hill in Warwickshire, the site of the first battle of the English Civil War in 1642. The London borough of Merton, SW19 has an Edge Hill district, while Edge Hill in Australia is the oldest and most picturesque suburb of Cairns in northern Queensland. Other Edge Hills are in Georgia, USA and Jamaica.

Elsie Arrives

From 1923, in the pages of the Edge Hill Magazines, a new person appears who became increasingly important to Edge Hill. Miss E M Butterworth, a University student from 1902 to 1905, 'came to live with us', and had 'been generous in giving the benefit of her special knowledge of Geography'. In a very short Principal's report in 1925, Miss Smith noted 'we welcome on the staff next session, Miss Elsie M. Butterworth, a past student, who is taking full responsibility of the subject of Geography, and will help with Education'. She had a varied and extensive teaching career and had studied at Oxford, Paris and Grenoble, in addition to Liverpool. Shortly after taking up a teaching post at Edge Hill, she collaborated with L W Lyde, Professor of Economic Geography at University College, London, in writing a series of textbooks, in a strongly imperial mode.[11] Lyde was a prolific writer of school texts and is said to have sold over four million. He had a profound influence on geography teaching in the first decade of the twentieth century (Balchin 1993:102). Foreign embassies complained about disparaging portrayals in British textbooks, for example in *The Netherlands*, Lyde claims that 'Frisians were the cleanest and most independent of the Dutch people', implying negative characteristics to those in other regions: Lyde was accused of stereotyping and mixing fact and opinion. Geographical determinism was also a major theme in his texts; people living in mountain areas were deemed independent, dull and steady living in plains, while Arctic climates 'stunted the brain' (Marsden 2001:137). How easily she collaborated with him is not recorded. Miss Butterworth liked fieldwork, but in 1901 Professor Lyde had opposed fieldwork as being too time consuming (Marsden 1976).

---

[11] E-mail Bill Marsden, January 2005

Miss Butterworth, Tutor in Geography, is publishing several books, in co-operation with Professor Lyde, suitable for Elementary schools. The series is the 'Why and Where Geographies' and is published by Blackie. 'From Pole to Pole' and 'The Overseas Empire' are already in circulation (1927) and 'The Old Country' has quite recently been published...First Years saw the Eclipse, several tutors, having journeyed by taxi to the neighbourhood of Southport, were greatly favoured in the view they obtained of the various phases of this astonishing spectacle. (Edge Hill College Magazine No. 34 1925)

In 1928 while waiting to move to the new site in Ormskirk Edge Hill's old building, 'owing to the dangerous state of the electrical system' had to be entirely rewired, 'and a very large sum of money spent on the old building'; it lacked adequate heating and was very cold in winter. In 1929 the Principal, Miss Smith, reported that the Lancashire Education Committee has 'bought a piece of land on the edge of Ormskirk', but it was to be four years before the actual move, during which time normal College life continued. The Principal gave a lecture on 'Some impressions of Italy', a Rambling Club formed which had 'many enjoyable rambles in the Wirral District' while Miss Deakin is mentioned as leading the violin accompaniment to the Carol Evening.

Coach Excursions

Miss Deakin had organised her geography excursions to local places, on the Wirral, using public transport (Fig 1.5) but Miss Butterworth favoured slightly more distant locations, particularly to the Lake District, and clearly liked 'coach tour' type of excursions; as a student of Edge Hill, Miss Butterworth had found 'long treks on foot or by tram' stressful. The GA had initiated 'conducted motor tours' in 1921; Miss Butterworth appears to be an early adopter of this form of transport. She also introduced visits to the docks. Accounts of these visits in the Edge Hill College Magazine are amazingly long, often whole page descriptions of the entire journey and activities, in flowery language, reflecting Miss Butterworth's writing style, as she became magazine editor in 1933. The coach journey was clearly expected to be part of the learning experience, and may have provided valuable opportunities for students with little access to transport to see other places. It is interesting to read of the route north from Liverpool, pre M6 motorway, and to appreciate the time taken to reach the Lake District in the early 1930s. The last geography excursion from Edge

Hill in Liverpool, in the summer of 1933, ambitiously marked a return to the use of public transport, by train, bus and steamer on Ullswater.

Fig 1.7 Aira Force, Lake District, Geography fieldwork excursion 1933

Source: Sylvia Woodhead

1929: The Geography Class visited Windermere. The view from Orrest Head was very clear, and the sail across the lake was a great delight.

1930: A lecture was given by Miss Butterworth on 'Mountains of Norway and those of Italy' by lantern slides. Miss Butterworth took a party of students to visit Messrs.

Cammell Laird's shipbuilding works at Birkenhead, where all found much of interest in their outing. First Year Students of Geography drove to Llangollen and back, a differing route being taken on each journey. Much interest was aroused by the different changes in geology formation to be observed on the outward journey via the Wirral, Wrexham and Ruabon varying from the Triassic in Wirral peninsula to the old Silurian of the hills west of Llangollen, the coal measures of Ruabon being an intermediate stage. At Llangollen a climb was undertaken which afforded a fine view of the magnificent limestone escarpment seen to the north of the synclinal valley, and some of the party also visited the Horseshoe Falls, returning via Valle Crucis Abbey. The return journey via Cefn, Wrexham and the Runcorn Transporter Bridge was greatly enjoyed, and all the students who took part in the expedition profited greatly thereby.

1931: Windermere was the lake visited by the students of Geography and this experience was enhanced in value and enjoyment by the fact that the journey was made by road and not by rail, Miss Butterworth and Miss Pope both being members of the party. Having reached Ambleside, the Kirkstone Pass was ascended from the top of which an excellent view was obtained of the district, including Brothers Water. After a short rest, during which observations made on the journey were discussed, as well as the geological structure of the neighbourhood, all returned to Bowness by way of the village of Troutbeck. Tea having been partaken of, the return journey was made via Kendal, Lancaster and Preston, and a very happy party alighted at the College gates as the clock struck ten. In addition Miss Butterworth also arranged a visit to a liner.

1932: The Geography expedition to Ambleside (with Miss Butterworth and Miss Pope) was taken in glorious weather, and as this to many of the party was their first visit to the Lake District, they were able to see that beautiful region under the best possible conditions. The comfortable buses in which the journey was made contributed also to its pleasure. The route lay first through the flat pasturage between Liverpool and Preston, and then through more changeful scenery. To the East of the road lay the moorlands and uplands of the Pennine Fells and to the West were the meadows and grasslands of the Fylde. The towns of Garstang and Lancaster were passed through en route and were followed by Carnforth and Kendal. The view of the Pennines was impressively beautiful, and the northern shore of Windermere was hailed with delight as the proximity of Ambleside was quickly realised. The buses were then deserted, and the occupants set out on foot to climb Lough Rigg, whence they obtained a delightful view of Lake Windermere, the Langdale Pikes and Scafell. They then walked past Grasmere and Rydal Water back to Ambleside. The intense green of the hills and the numerous trees gave the impression of well wooded country, and were most refreshing to the eyes of the city dwellers. The drive back to Liverpool was made all the more enjoyable because a detour was made which provided pleasurable variety. In addition to the above expedition, a visit to a liner was arranged by Miss Butterworth for June 23, when the students had the pleasure of seeing the magnificent White Star Liner, MV 'Georgic', the sister to MV Britannic, the arrangements of which filled all its visitors with surprise and wonderment, and the examination of which most happily occupied a whole afternoon.

1933: We travelled for Penrith by bus to Pooley Bridge, and thence by steamer along Ullswater, passing first low-lying wooded shores, until the middle reaches, the scenery became bolder, and the mountains rose sheer out of the lake. On arrival at Patterdale, Miss Butterworth conducted a party along Grisedale, whence Helvellyn could be seen in the distance. Meanwhile Miss Pope took another party to Aira Force, where the falls were admired from the lower and upper reaches. The return was made by steamer and bus to Penrith, where the whole entrained for Edge Hill, having spent a happy day in the beautiful Lake District (Fig 1.7).

The White Star Line

The comfortable, luxurious ships named the Oceanic, Atlantic, Baltic and Republic, were launched in 1870, the first steamships of the White Star Line. All White Star liners had names ending in –ic; the most famous, the Titanic, was launched in 1912. The Georgic was the last liner to be built for White Star before it became part of Cunard. The 2-funnel liner with a top speed of 18 knots was launched in 1931. Edge Hill students led by Miss Butterworth visited the ship just two days before its maiden voyage from Liverpool to New York on 25 June 1932. In the Second World War, the Georgic was initially converted to a troop ship and evacuated troops from Norway, and was bombed at Suez in 1941.

Later Fate

In 1930 in the College Magazine it was noted that plans for the new College had been seen:

I think we all agree that we may expect a beautiful imposing building. There will be an Education Block and four hostels, all parts of the College being reached by covered ways. There will be grass quadrangles, the main one being bounded by the Education Block, the Dining Hall and the Gymnasium. The Assembly Hall will be panelled and decorated with the Building Fund money in memory of Miss Hale. Threw will be stage and a gallery, altogether giving seating accommodation for 450 people. Although there will be four Common Rooms and each Hostel will develop some individuality of its own, we hope to preserve the College feeling of unity, through the use of one Dining Hall and the life of all the various Societies which are now so vigorous. (Edge Hill College Magazine No. 39 1930)

In 1933 there was an impressive gathering of the Edge Hill Guild of former students, an occasion tinged 'with its own sadness', because the following year 'we shall be living in different surroundings, and to some extent a different life'.

We had to face the different way of life in the country as compared with our own very urban existence in Liverpool. Our very extensive connection with Girls' Clubs would be impossible, but Games and Swimming would need no travelling. We would miss the many interesting lectures the city provides but country pursuits would add something to the lives of those students who lived in towns… Most of the leaving students felt a strong sense of loyalty and all of us were aware that much history had been made there and that many hundreds of students had spent very happy years within its walls. (Miss E M Smith, former Principal, writing in Edge Hill College Magazine No. 63 1954)

The Durning Road premises were eventually purchased by Liverpool Education Department, and became a Junior Technical School. The large underground shelter in the basements of the Durning Road building sustained a direct hit in the wartime blitz of Liverpool, on 29 November 1940, when about 300 people were packed into the basement. A mine hit the building which then collapsed into the shelter below, crushing many of the people sheltering there. Boiling water also poured in from the fractured heating system. 166 people were killed and many more were injured in this worst single incident in Liverpool during the Second World War. The ruined building was later demolished, and a 1957 OS map of the area shows its location as blank. In 1958, a Boys Secondary Modern school of modern design was built on the site. Called Edge Hill

Secondary Modern Boys School to preserve the name and links, the school was presented on its opening with a picture by the then Principal of Edge Hill, Miss Margaret Bain[12]. Tom Eason, writing a history of Edge Hill in the 1965 Edge Hill College magazine, comments that a living sense of continuity was broken with the move from Durning Road, and deepened by the tragic destruction of the building in 1940. He was concerned that 1960s Edge Hill students did not know the Hale connection, and wondered if some of the traditions and strengths of a town based college had been lost in the move to Ormskirk. However in 1932 Miss Smith referred to 'the present building which was proving inadequate for the purpose for which it was intended, owing to the rapid increase of the industrial area in which it was situated'.

Despite its cramped conditions, the Durning Road site did have all the facilities and culture of the city of Liverpool easily accessible. During the period from 1885 to 1933, Liverpool was a major trading port, the second in the British Empire, a world market place for cotton and grain, and a gateway to the New World. Walton, Wavertree, Toxteth and West Derby were fully incorporated into the city for the 1901 Census of 716,000. Bootle for example developed from a rural township into an incorporated borough with a population of 58,000 in 1901 (Muir 1907:298). Dominated by the docks there was also an important shipbuilding industry, iron foundries, glass and soap making. It was experiencing a great expansion of housing, learning (Liverpool University predates Edge Hill by only three years) enterprise and energy. The Liver Building was built in 1911; Liverpool was at a high in the 1930s. From 1885 to 1933 Liverpool had shown considerable growth, unlike Ormskirk's development, recounted next. Transition from urban Liverpool to rural Ormskirk could not have been more contrasting. Miss Smith mentioned 'unspoken fears about isolation' (Montgomery 1997:46). The next section looks at the rural background to Ormskirk up to the 1930s, when the College moved from the town to the countryside near Ormskirk.

---

[12] Edge Hill College Magazine, No 67 1958

2: Ormskirk before Edge Hill

In the earliest geographical traditions of exploration, this chapter uses accounts of early travellers to Lancashire, plus extracts from written histories[1] to establish Ormskirk as an agricultural market town, built around a cross roads, whose urban morphology still retains a compact centripetal pattern. It examines the character of Ormskirk before the College moved from the Edge Hill district of Liverpool, while the nature of the College site in the countryside outside Ormskirk is researched from tithe records and maps.

Early Travellers

In 1933 Edge Hill moved to a very different location, on a countryside site south east of Ormskirk, a small, distinctive market town, in a remote corner of North West England, between Liverpool, Preston and Wigan (Fig 2.1). Ormskirk's development contrasts very much with that of Liverpool, as can be seen by looking at the accounts of early travellers to Lancashire. One of the earliest to visit Liverpool was **John Leland** in 1533, commissioned by Henry VIII to visit libraries and antiquities of England.

> Lyrpole, alias Lyverpole, a paved town hath but a chapel. Walton four miles off not far from the sea is the parish church. The king has a castlelet here and the Earl of Derby hath a stone house. Irish merchants come much thither as to a good haven. Good merchandise at Lyrpole and much Irish yarn that Manchester men do buy there.

William Camden in 1582 in his journey through England, using Leland's notes[2], merely recorded that Ormskirk was a market town famous as a burial place for the Stanleys. **Daniel Defoe** travelled through Lancashire in 1690, and was impressed by Liverpool, calling it one of the wonders of Britain. From Liverpool, Defoe travelled to Warrington and Manchester, then Bolton. Defoe recounts, in some detail, the famous siege of Lathom House in 1644, when the Countess of Derby defended the house. Defoe considered the connection with

---

[1] Padfield (1986) writes probably the most accessible history of Ormskirk. Duggan (1998) documented Ormskirk's growth from 1660-1800, while Stacey (1962) described nineteenth century Ormskirk from manuscript sources supplemented by reports in the local newspaper, the Ormskirk Advertiser, founded in 1853
[2] William Camden's *Britannia*, written in Latin, was based on notes on earlier journeys by John Leland

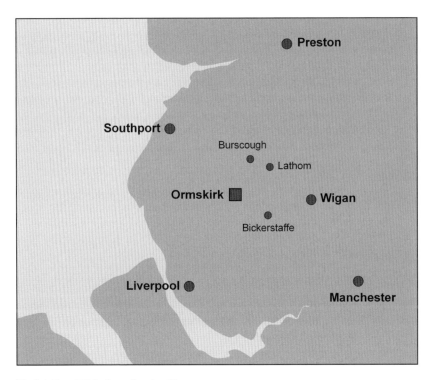

Fig 2.1 Ormskirk in its regional setting

Map: Ann Chapman

the Stanley family was the most important aspect of Ormskirk. His journey continued through Preston, which he thought a fine town.

> There is no town in England, London excepted, that can equal Liverpool for the fineness of its streets, and the beauty of the buildings... On the left hand of this town, west, even to the sea-shore, there are not many towns of note, except Wiggan on the high road, and Ormskirk, near where we saw Latham House... We saw nothing remarkable in Ormskirk, but the monuments of the ancient family of the Stanly's, before they came to the title of Earls of Derby (Daniel Defoe, 1690).

The vivid journal of **Celia Fiennes**, an early traveller through Britain between 1685-1712, provides a comprehensive survey of Britain. On horseback and with only two servants, she undertook a large number of journeys; she entered every English county, and crossed into Scotland and Wales. In addition to stately homes, she was also interested in crops, drainage of the fens, mining; rock salt in

Cheshire and fluorspar in Derbyshire, and manufacturing. She was interested in spa or mineral water for 'natural cures', perhaps against the dominantly meat and wheat diet of the wealthy. She was an authority on spas, and had theories as to the precise healing properties of the water. Obviously robust in health, she was a connoisseur of beer, the universal beverage. She shared the taste for neat, artificial gardens, 'neat' being her greatest praise. Not as well educated as men at that time, her spelling was wildly eccentric and punctuation was just a few commas. Unlike some other travellers, Celia Fiennes was independent, describing only what she saw at first hand. Despite being interested in the attempted drainage of Martin Mere, she didn't travel to Ormskirk as part of her journey to Liverpool in 1698. She travelled to Prescot, passing by Knowsley the Earl of Derby's house, by lanes to Wigan. She comments that she passed a mere.

> [Liverpool is] A very rich trading town, the houses of brick and stone built high and even, …its London in miniature (as much as ever I saw any thing)… But not going through Ormskirk I avoided going by the famous Mer call'd Martin Mer …it being nearer evening and not getting a Guide I was a little afraid to go that way it being very hazardous for strangers to pass by it (Celia Fiennes, 1698, in Morris 1982)

Arthur Young, a farmer and writer of books on agriculture, toured northern England in 1770. He advertised his journeys in advance, and collected letters of introduction and information about which leading estates and farms were worth visiting. He detailed crops, livestock, practices, costs and profits, ideally from the farmer, though he had no faith in the wisdom of small farmers. He described the general system of farming of an area. He is also well known for his strong views on the poor roads of the time. The following account of the road to Wigan is typical:

> To Wigan. Turnpike. Very bad. I know not in the whole range of language terms sufficiently expressive to describe this infernal road. To look over a map and perceive that it is a principal one, not only to some towns, but even whole counties, one would naturally conclude it to be at least decent; but let me most seriously caution all travellers who may accidentally purpose to travel this terrible country to avoid it as they would the devil, for a thousand to one but they break their necks or limbs by overthrow or breakings down. They will meet here with ruts, which I actually measured four feet deep, and floating in mud only from a wet summer; what therefore must it be after a winter? The only mending it in places receives is the

tumbling in some loose stones which serve no other purpose but jolting a carriage in the most intolerable manner. These are not merely opinions but facts, for I actually passed three carts broken down in these 18 miles of execrable memory. (*Northern Tour* (1770) IV 576-81, quoted in Mingay 1975:156)

Only Nathaniel Hawthorne, US consul to Liverpool from 1853 to 1857 described Ormskirk market:

From the church, a street leads to the market place, in which I found a throng of men and women, it being market day; wares of various kinds, tin, earthen, and cloth, set out on the pavements; droves of pigs, ducks and fowls; baskets of eggs; and a man selling quack medicines… The aspect of the crowd was very English; portly and ruddy women; yeomen with small clothes; broad brimmed hats, all very quiet and heavy, good humoured, and many of them, no doubt, boozy. I could not easily understand more than here and there a word. (Stacey 1962:3).

Ormskirk was seen by these early travellers as an ancient market town associated with adjacent aristocracy. Its location in the centre of an agricultural plain has none of the advantages of Liverpool. Its market, fair and small-scale craft industries supported local agriculture, in a similar manner to other English country towns. Ormskirk's misfortune is perhaps to be surrounded by settlements, Wigan, Southport and Preston and of course Liverpool, which have all become larger. Ormskirk is not on the coast, so could not develop into a port. Neither is it on a navigable river, bringing trade; lying on small watershed of the Douglas to the north and Crossens to the south, its streams are small and indeterminate (Fig 2.1). Small scale handloom weaving developed, but Ormskirk was always on the edge of the Lancashire cotton industry. It developed a small-scale silk industry; an entrepreneur from Macclesfield saw Ormskirk as providing cheap labour. It was avoided by the Leeds to Liverpool canal, which approaches only to Burscough three miles to the west. Ormskirk has no resources of coal, metals or good quality building stone. The underlying rocks are New Red Sandstone, which is quarried only for local building stone. The nearest collieries of the Lancashire coalfield were at neighbouring Bickerstaffe in 1871 twenty-four Ormskirk colliers walked to work there (Stacey 1962:78). Any raw materials to support industrial development in Ormskirk had to be transported in, and poor transport was always a problem in the past. So while Liverpool developed into a major port, Ormskirk grew much more slowly. It has retained a completely different, rural, character; its most famous product is potatoes, about which little is written. Its spa development failed, the Derby family left, and by 1933 Ormskirk had reverted to its market town function.

Origins

The **early settlement of Ormskirk** evolved on a sandstone ridge at the western end of the Skelmersdale Plateau at the southwest edge of the South Lancashire coalfield, in a dominant position overlooking the South West Lancashire Plain. Until late in the eighteenth century, this plain was undrained mossland, extending to the coast and surrounding a large shallow lake, Martin Mere near Rufford to the north of Ormskirk. Although itself above flood level, the surrounding mosslands seriously deterred Ormskirk's development and visitors and its poor transport hindered trade. When Martin Mere was drained for crop growing, it provided Grade One agricultural land; Padfield (1986:75) argues that Ormskirk benefited from Liverpool's growth by providing carrots and potatoes. The River Douglas flowing north towards the Ribble estuary provided an often impassable barrier between Ormskirk and Preston, while to the south the River Alt was also surrounded by marshy ground. The Skelmersdale plateau provided access from East Lancashire; Ormskirk developed as a route centre in West Lancashire, where east-west and north-south routes converged. Its ancient urban fabric is essentially a cross near a church; to the west is Church Street, Burscough Street to the north, with Moor Street to the east and Aughton Street to the south.

The **name** Ormskirk is of Scandinavian origin. *Orm* is believed to be a personal name, possibly an early lord of Lathom (the parish east of Ormskirk) while *kirk* is Scandinavian for church. Ormskirk's church is an integral part of the town's development, and probably developed before 1189, though the church was not mentioned in the Domesday survey of 1086. Padfield (1986:16) visualises Ormskirk developing from the church to the crossroads. In 1198 nearby Burscough Priory confirmed the foundation charter of Ormskirk church, and in 1286 Edward 1 granted the rights of a weekly (Thursday) market and annual five day fair in August. In 1461 Edward IV granted the right to hold an additional Whitsun fair, as trading had increased. In 1598 Ormskirk was described in the duchy court as a 'great, ancient and very populous town, and the inhabitants are very many, and a great market is kept there weekly besides two fairs every year, and quarter sessions are held there twice a year whereunto… great multitudes of people continually thither repair' (Duggan 1998:xvii).

Since 1603 when the township of Ormskirk was granted to William Stanley, Sixth Earl of Derby, Ormskirk's fortunes have been tied to the **Stanley family**. Their family home at Lathom House was twice besieged in the Civil War, and in 1645 Ormskirk's inhabitants suffered as soldiers from both sides obtained food and supplies legally and by plunder. Malnutrition and presence of the military led to a devastating outbreak of plague in 1648; town and market were closed to avoid spreading of infection, and many Ormskirk residents suffered extreme poverty for up to five years. Lathom Hall has now been so thoroughly demolished that no certain knowledge remains of its site.

Ormskirk was a short-lived **spa** town and leisure resort, which did not prove successful. In 1670 iron-rich spring water at Spa Farm, Lathom, a mile and a half east of Ormskirk, was said to have healing properties and a small summer spa developed for which Ormskirk provided accommodation and services for visitors. The well was walled in freestone with seats and a wooden roof, while nearby walks were provided, following the fashion first developed in Tunbridge Wells. A cold plunge bath was built in Ormskirk, but it never became fashionable as a spa, and by 1720 had fallen into disuse, as being too remote. The Earl of Derby sealed Ormskirk's fate in abandoning the traditional family home at Lathom and moving to Knowsley. **Ormskirk Races**, possibly held along Long Lane, were described in 1685 as an established sporting venue with an enthusiastic following but by the late eighteenth century were gone, again as consequence of withdrawal of Derby interest. Focus moved to Knowsley as the Derby family moved their famous races (Duggan 1998:169).

Fair And Market

Ormskirk Fair started as a Horse Fair, where animals such as horned cattle, horses, sheep, oxen and pigs were sold. Pottery souvenirs, rugs, gingerbread and later clothes and linen were also on sale. In 1741 Richard Lathom of Scarisbrick bought gingerbread and in 1744 a rug (Duggan 1998:4). The Fair was also an occasion for entertainment; in 1714 there was a tiger, dancing with swords, juggling and a musician, and for socialising; Nicholas Blundell's diary mentions drinking with friends in the Lamb (1706) The Golden Lion (1715) and the Talbot (1725) (Duggan 1998:5).

Duggan (1998) documents how the local produce sold at **Ormskirk market** changed over time, from corn to dairy products, fish to potatoes, then to livestock, at least partly from increased competition from the growing town of Liverpool. In 1754 Ormskirk was noted as a great **corn market,** where barley, wheat, oats and rye were sold, though by 1798 very little grain was sold. Barley was important for Ormskirk's brewing industry, while oatmeal was a basic food for poorer people, sold in the meal house, which was included in plans for rebuilding the Town Hall in 1779. Grass was sold at the cross, where sellers had to demonstrate a legal right to sell it, and peas and beans were sold wholesale. By 1750 fowl, butter and eggs were sold, and often resold in Liverpool on Saturday. Fish, very difficult to keep in those times when quick sales were essential, was also traded at Ormskirk market. Salmon was popular, but eels caught in Martin Mere and coastal cockles, mussels and shrimps were also available. By 1794 Ormskirk fishmongers were dealing in large quantities of fish, while by contrast little fruit was sold. **Potatoes**, grown locally on sandy loam soil, were sold on Ormskirk market from the seventeenth century. Horse sales, of horses, geldings, mares, and colts, expanded from the fair into the market before 1660, though they were restricted to the east end of Moor Street, away from the prevailing wind, while the pig market was held behind the houses in Moor Street to keep disturbance and dirt from the market centre (Duggan 1998:11). Trade in cows and pigs continued to the nineteenth century. There was a thriving **butchery trade** beneath the Town Hall, with extra stalls on the Thursday market. The shops were often very dirty; meat was often sold on credit with the butchers' debts not paid. The County Court, established in 1841, records local farmers selling commodities like butter underweight, while in their turn farmers complained that Ormskirk failed to provide a covered market. The Ormskirk Advertiser of 1856 commented that the annual cattle and horse fairs continued to attract purchasers of local or Irish and Scottish cattle fattened locally (Stacey 1962).

Besides the market and fairs Ormskirk had some **handloom weaving** and small scale manufacturing, none of which now exist. Craftsmen, potters and basket makers, prominent in the seventeenth century disappeared from records in the eighteenth century. The woollen industry had a long history in Ormskirk; Norfolk wool was hand spun by women. **Soap** manufacture, made by mixing tallow, oil or fat with potash derived by burning bracken, developed to supply the linen trade. The earliest soap boiler was at the back of a house in Church Street; it was not a lucrative trade and eventually lost out to soap made at chemical works on the Mersey. **Linen,** sold at Ormskirk Fair, was used for sailcloth for Liverpool shipping, but its manufacture moved to Warrington (Duggan 1998:36). Fustian, finer cloth with a linen warp and cotton weft, was recorded in Ormskirk in 1770s. An 1830 gazetteer records Ormskirk's chief trade as **cotton** spinning, and coarse thread made for sailcloth, also ropes and some business in silk. In 1894 the Ormskirk Advertiser claimed that considerable business in spinning cotton for Manchester manufacturers took place in Ormskirk, though the 1841 Census records only seven cotton workers out of 300 weavers and winders; 224 working in silk, and eighty-four unspecified. Around 1825 emphasis seems to have switched to **silk** weaving; silk agents are named in an 1854 directory, acting for Samuel Barton of Macclesfield, with three silk manufacturers, in Burscough Street, Aughton Street and in Church Alley. The 1851 census records over 300 silk weavers in Ormskirk with nine recorded as factory workers, though by 1861 only 174 silk weavers were named and no factory was mentioned. The Ormskirk Advertiser of 1855 noted that nearly all silk weavers were out of work owing to the scarcity of work, and one mill had closed, another's work was limited. In 1894 the Advertiser reported that the sound of shuttle was very rarely heard. The decline of handloom weaving continued and several schemes to build cotton mills in Ormskirk failed (Stacey 1962:7). Ormskirk suffered in adopting handloom weaving at a time when it was already declining in the face of extension of power loom weaving. It was also on the fringe of the cotton area, and far removed from the Macclesfield silk area. It appears that manufacturers used Ormskirk as a source of cheap labour in busy periods. (Stacey 1962:36).

Ropes were made from hemp, which was grown locally; hemp yards are recorded on the early deeds of Ormskirk properties. The preparation involved 'retting' or soaking hemp fibres for two to three weeks till rotted, which rendered the water noxious. Around 1850, a major Ormskirk roper employed sixteen men and twenty-seven boys and apprentices. In 1676 a scheme to tap the potential of imports from American colonies foundered; a small **sugar refinery** was started in a kitchen in Bickerstaffe, but was abandoned after three years (Duggan 1998:17). Ormskirk long case clocks formerly had a high reputation (Duggan 1998:40).

Georgian Luxury

The period from 1730 to 1777 was possibly the hey-day of Ormskirk, when rich people lived luxury lifestyles, though few Georgian town houses remain as evidence. Trade extended from Ormskirk market into shops; one trader advertised himself in 1750 as bookseller and ironmonger, selling various pills. Specialist occupations, such as cabinetmaker, organist and brandy merchant, show the luxurious lifestyle of the affluent of the Georgian era. In 1777 Ormskirk people were said to quickly copy London fashions (Duggan 1998:26). Doctors, apothecaries, teachers and lawyers were high in Ormskirk's social hierarchy, but their numbers fell dramatically from 1750 to 1800 as they left for larger towns, either Wigan or Liverpool (Duggan 1998:29). As professional people left to progress, Ormskirk's workers reflected a market centre meeting local needs. Weaving was replaced by other occupations, such as builders; joiners and carpenters, bricklayers, paviours and flaggers, stonemasons and quarrymen, plasterers and painters, plumbers and glaziers and sawyers. Other craftsmen were recorded in smaller numbers; shoemakers, tailors, cabinet makers, nailers, twine spinners, metal workers, hatters, watch and clock makers, dyers and upholsterers (Stacey 1962:8).

Many of Ormskirk's population around 1850 were employed in **personal services**, as dressmakers, washerwomen, manglers and charwomen, innkeepers, hostlers and grooms, makers of stays, corsets, bonnet and caps, professors of music and dancing, while many were domestic servants. There were also distributive trades; many butchers, grocers, carters, railway and postal workers, and small number of professional (men) in law, medicine, the church, banking, civil engineering, poor law administration and civil service, with many governesses and schoolteachers. All illustrate the very small size of most commercial concerns in Ormskirk at this time; most employers had very few workers. From 1870 to 1900 newspaper editorials gave the impression of industrial and commercial stagnation in Ormskirk, with people investing in industrial shares elsewhere. (Stacey 1962:11) Unlike Liverpool, Ormskirk was only partly and impermanently industrialised.

People gained from Ormskirk's market activities by providing medical, legal and commercial **services**, acting as bankers and providing agricultural machinery. Census returns show that a number of farmers continued to work small plots from their town houses in Ormskirk, some innkeepers kept prize boars and stallions at stud, and nearly 100 farm labourers lived in the town. Others acted as dealers in potatoes and pigs, although by 1879 wholesale merchants in Liverpool had a powerful hold on trade (Stacey 1962:5). The 1851 Census shows Ormskirk occupations, such as millers or butchers, dependent on local agriculture. Others were more concerned with services to farmers, while an agricultural mechanic was first recorded in the 1861 census. Reference is made at intervals to the limited scale of trade and manufacturing in Ormskirk. A 1887 directory mentions only two wholesale brewers and three hat manufacturers, which once claimed to supply most Lancashire towns. (Stacey 1962:5)

Ormskirk's trade development was inhibited by **poor transport** (Duggan 1998:47) and highways were recorded, as for example by Arthur Young in 1770, as impassable. Most goods for market were carried on horseback as roads were too bad for coaches. As early as 1746 narrow bridges and roads created bottlenecks on market days, and were widened to accommodate the traffic and market. The improved Preston to Liverpool Turnpike Road of 1771 led to an increase in carriers and coaches and greater demand for local inns, but also led trade away from the town, and more of the community followed professional people and moved to Liverpool. The **railway** from Liverpool via Ormskirk to Preston in 1849 allowed excursion visitors, while Ormskirk was described as a desirable place to live (Padfield 1986:94) but also allowed Ormskirk people to travel elsewhere; improved transport has always been a two edged advantage.

Ormskirk's early **buildings** were a mixture of black and white timbered, stone, lime washed wattle and daub, brick or stone; there was no unity of urban design (Padfield 1986:27). Wattle and daub houses were replaced by bricks, though, in 1754, some old properties still had clay walls and thatched roofs. **Bricks** were locally made (Duggan 1998:67) from locally dug clay fired with local coals, but the industry moved to Burscough. 1830 and 1850 directories describe Ormskirk town as well built, though of rather ancient appearance; Court records show that considerable repairs were needed; roofs were poorly thatched or slated, window shutters required attention and chimneys were dangerous. Changes needed approval of the Earl of Derby, who proved reluctant to release land for housing with the consequence that infilling led to **courtyard** developments. These enclosed yards in the long properties going back from the road became overcrowded slum dwellings, though the frontages appeared unchanged.

Slow Growth

Ormskirk's population grew slowly largely by immigration; census records of 1811 to 1815 show baptisms exceeded deaths by about twenty. Immigration was from surrounding villages; Rufford, Bickerstaffe, Halsall and Scarisbrick, bordering counties; Westmorland, Cumberland, Yorkshire and Cheshire as well as from Ireland. An **Irish** presence in Ormskirk was first recorded in 1834, with an influx of refugees fleeing the Irish potato famine from 1845 to 1850. Most were farm labourers, both men and women, aged between twenty and forty, who came as temporary harvesters, until as late as

1887. Others were employed as pedlars, hawkers, dealers in herrings, pipers and fiddlers, pig dealers, and stonemasons. A large number came from County Mayo (Stacey 1962:11).

There were always fears of epidemics in early Ormskirk. **Cholera** struck in 1849 resulting in high mortality, which was linked to poor conditions in overcrowded courtyard dwellings, which had few pumps, wells or privies, while sewage flowed down Aughton Street. Concern over the condition of lodging houses and frequent cases of typhus led to the establishment of the Local Board of Health; Ormskirk was one of the first towns to have a Board of Health as a result of the Public Health Act 1848. Records show a slow start, with considerable ignorance of public health laws and inability to keep satisfactory accounts; the Board of Health was massively criticised in the local press. Accusations and recriminations dominated local politics, and costly legal altercations occurred in the 1860s when tenants of Lady Scarisbrick complained of Ormskirk effluent polluting their watercourses while in 1870 Aughton Street was described as covered in rotting fish entrails. When in 1894 Ormskirk Urban District Council was created, local political wrangles still continued (Stacey 1962:21).

Before 1848 Ormskirk was dependent for water on scattered public and private wells sunk into the saturated sandstone of the New Red Sandstone formation, most of which were subject to surface contamination from middens, drains, cesspools or churchyards. There were over two hundred wells, but only about ten had water fit to drink: over one hundred people were counted waiting for water at one pump (Ormskirk UDC 1948). Amazingly there was some opposition to the provision of a piped **water supply** via a reservoir and water mains system that was eventually set up in 1853, with a sewage system installed in 1854 (Duggan 1998: 190) by which time seven hundred houses had public water supply. A small cottage hospital opened in 1896 to avoid delays in travelling to Liverpool for treatment. Ormskirk was one of the first towns to have gas lighting, in 1835; it was quite a go-ahead town then. However because of political wrangles, electricity was not installed until 1912, in Ormskirk parish church, compared to 1896 at Durning Road, Liverpool.

Until 1875 Ormskirk was subject to the manorial rights of the Earl of Derby, exercised through his steward; inhabitants were fined if they failed to attend the annual Court, though they became increasingly reluctant to do this. In addition to the steward the Court also had twelve jurors and many officials; Church Warden, Constable, bailiffs, clerks, inspectors, ale tasters, supervisors and engineers. By

1836 many of these officers were as superfluous as the curfew bell, and the stocks and ducking stool recently removed. Until 1875, the Court kept the town and market in order; it punished those dealing in short measure, checked and adjusted the placing of market stalls, it imposed fines for medieval offences, and checked the sweeping of roads in front of houses, enforced ringing of pigs and muzzling of dogs (Stacey 1962:14).

In the 1860s the police took over some Court functions. They dealt with routine felonies and misdemeanours such as stone throwing, belt fighting, cruelty to animals, fights between local and Irish farm labourers, possession of unlicensed dogs, playing football in the streets, accompanied by atrocious language and working on Sundays. They also had the duty of inspecting lodging houses to check on overcrowding. The police were preoccupied with the effects of intemperance. In 1876 one licensed house existed in Ormskirk for 127 inhabitants, and over 200 prosecutions for drunkenness were sustained. In 1886 fifteen per cent of the properties in Aughton Street were devoted to the consumption of alcohol. In 1887 there was one licensed house for each 147 inhabitants, more than in Birkdale or Formby (Stacey 1962:18).

Ormskirk gradually reverted to a small rural settlement on the remote West Lancashire Plain, where many enterprises, silk milling, sugar refining, a spa and leisure town, had developed and failed. The Derby family ceased to use the family vault in Ormskirk church, sold the Lathom House, and sold or gave away property for schools, churches and playgrounds, and in 1874 sold the market tolls in Ormskirk thus ending their connection with the town. As the influence of the Lord Derby's steward declined, Ormskirk fell into middle class control, its professional and entrepreneurial classes largely left for Liverpool, leaving Ormskirk devoid of innovative leadership (Stacey 1962:82).

In 1886, when Edge Hill was developing in Liverpool, an article quoted in the Advertiser reported Ormskirk as a quiet little town known for the strength of its rope and excellence of its gingerbread. Its apparent stagnation was repeatedly noted; it failed to attract major industries. Neither were its tastes very highbrow: hare coursing and pigeon and sparrow shooting were popular recreations noted in the Advertiser. The athletics club lasted only from 1867 to 1878, while educational classes and scientific talks attracted only small numbers (Stacey 1962:75). Galas, such as in 1894, exhibited local trades, with a historical pageant of the siege of Lathom; a harking back to a significant point in Ormskirk's history.

In the period 1885 to 1933 while Edge Hill was in Liverpool, Ormskirk had lost its corn mill, destroyed by fire in 1912, its workhouse ceased to exist in 1915, and its cattle fair and Corn Exchange stopped. Ormskirk acted as a junction for coal trains from the Skelmersdale area; it was a major railway depot for potatoes, and in 1885 was still trying to get parliamentary assent for a branch line to Halsall and Shirdley Hill. The first motor cars and motor buses found its streets narrow bottlenecks, and a by-pass was built in the 1920s.

Agriculture remained the mainstay and the town was a popular place for farmers' retirement. Many Ormskirk residents moved from the older town houses to newer properties built on St Helens Road or Ruff Lane, or moved to Southport. Thus Ormskirk welcomed the construction of a new higher education college on open land outside the town.

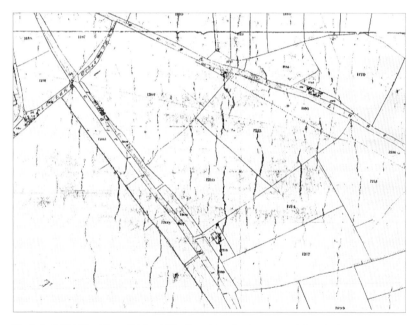

Fig 2.2 1845 Tithe Map of Edge Hill's Ormskirk site, from an original in the Lancashire Record Office. Between Rough Lane and Scarth Hill Road lie a few large fields.

Source: Lancashire Record Office

College Site

The college site, as it was before 1933, can be recreated by examination of maps and tithe records[3]. The **1845 Tithe map**, 'drawn for the payers of Ormskirk township' (Fig 2.2) shows Ormskirk town including gardens making up only eight per cent of the township, the rest of which was made up of seventy per cent meadow and pasture, sixteen per cent arable and five per cent wasteland. Closer examination of the Tithe map shows the campus site in 1845 as farmland bounded by Rough Lane (now Ruff Lane) to the north, and Scarth Hill Lane (now St Helens Road) to the south. Across Scarth Hill Lane lay the Township of Bickerstaffe, with a line of boundary stones shown along the road. Around sixteen fields, numbered 1210 to 1239 are shown on what is now the site of the campus. Rough Farm[4] and Woodlands Farm are shown along Rough Lane, with Slack House Cottages and Mount Pleasant along Scarth Hill Lane.

Woodlands Farm, Ormskirk

A picture of Woodlands Farm, where Edge Hill's Ormskirk campus now stands, can be gained from the 1910 Tithe schedule[5], which describes the farm 'with frontages to Ruff Lane and St Helens Road'. **The Woodlands** (a mansion) 'a large well proportioned stone and slated house in good condition and repair', lay on the opposite side of Ruff Lane, while the farm house and farm buildings were on the Edge Hill site. **Woodlands Farm House** was 'brick built with grey slate roof, an old building but in good condition and repair, with a parlour, kitchen, scullery, pantry and four bedrooms' **Woodlands Farm buildings** are 'old, in fair repair, and not very suitable for the present method of farming in the district'. The Farm outbuildings included 'two small store rooms, boiler house, barn, engine house, loosebox,

---

[3] **Tithes**, originally one tenth of the crop, payable to the Church, and later as money to landlords, were considered by Arthur Young to be a great burden on agriculture However the survey of England and Wales for the Tithe Commutation Act of 1836 provides a detailed survey of rural land use, and can establish the original nature of the College site. For each district three documents were prepared, a file, an apportionment and a map. Some Tithe files (PRO IR 18) include reports on the local landscape, farming practices and agriculture. The apportionment (PRO IR 29) lists the name of the owner and occupier, plus the state of cultivation. It includes a list of landowners. (microfilm at PRO 559) (Rawding 2001)
[4] Note that Rough Lane has changed to Ruff Lane over time
[5] Searches by Charles Rawding at the PRO in London on 12 November 2004 Woodlands Farm is shown on the 1910 map IR 133/S/220, and described in IR 58 77304 Schedule 658

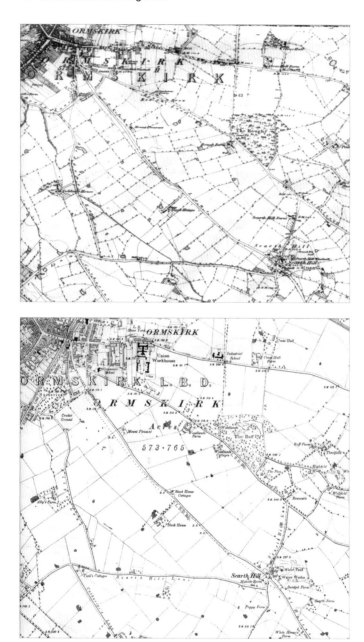

Fig 2.3 Edge Hill's Ormskirk site: Reproduced from OS 6" First Edition 1845 Lancashire sheet 92 (top)

Fig 2.4. Woodlands Farm in 1892 reproduced from OS 6" Lancashire sheet XCII.NW second edition (bottom)

Source: NGAS Map Collection

trap house and four loose boxes. Also a shippon for 20 cows, brick built with grey slates, 6 pigsties, and a Dutch barn[6] with 4 lays. The total area of the farm was given as 148 acres, 1 rood and 30 perches[7], and the total valuation (including the shooting rights) was £9605.0.0. The occupier was James Berry and the owner the Earl of Derby'. The Tithe description suggests a rather old fashioned mixed farm, with

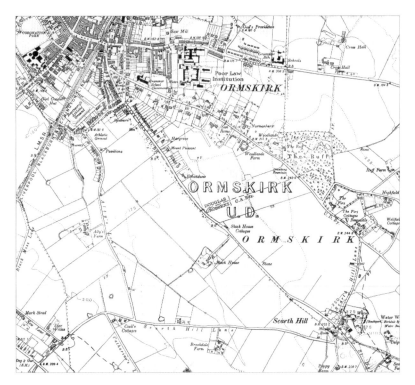

Fig 2.5 The St Helens Road site as acquired by Lancashire County Council. Reproduced from the 1927 OS 6" Lancashire sheet XCII.NW

Source: NGAS Map Collection

both crops and animals, at a time when it can be assumed that the district was turning towards more specialised crop production as occurs today, when the Ormskirk area is an important vegetable growing area.

---

[6] A Dutch barn has a curved roof set on an open frame and is used to store hay
[7] There are four roods per acre, and forty perches per rood. These units of land measurement were used until metrication

The first edition Ordnance Survey six-inch map of 1845, sheet 92 (Fig 2.3) shows Ruff Wood as The Roughs, with a benchmark of 257 feet, and a sand pit, opposite which Rough Farm is shown with two wells. Symbols suggest field boundaries of hedges and several fields contained ponds, possibly excavated for marl, to improve the soil. On the 1892 six-inch second edition map (Fig 2.4) the only buildings along St Helens Road (now named as such) are Mount Pleasant and Slack House Cottages. Rough Farm has become Woodlands Cottages, while the fields have halved in number and doubled in size. The 1927 OS six-inch map (Fig 2.5) shows continuous ribbon development of houses built out from Ormskirk along St Helens Road, past the house called Mount Pleasant, to The Hollies and Ethandune,[8] and along both sides of Ruff Lane, including San Remo, with Levens adjoining Woodlands Farm. Footpaths are shown from The Hollies to Woodlands Farm then south. Herbert Alcock at Woodlands farm was cousin of John Alcock at Slack House Farm over the road from the College. His son John still farms 'corn and cabbages'[9] there. Both farms were arable, but kept some pigs.[10] This was the site, on open farmland south east of Ormskirk, as acquired,[11] from the Derby Estate by Lancashire Education Committee. The story of Edge Hill's buildings at Ormskirk continues next.

---

[8] Now Edge Hill enterprises and history respectively
[9] Telephone interview with John Alcock, 23 March 2005
[10] Interview with Derek Sumner, August 2004
[11] Did the College buy the land, or lease it from the Derby Estate? The local view is that the Derby Estate is in general reluctant to sell land: certainly farmers like Richard Holland at Alty's Farm opposite College are still Derby tenants

3: A Campus in Ormskirk 1933-1946

This Green Campus 1933-1939

Having established the essential differences between urban Liverpool and rural Ormskirk, the focus now turns to pride in the new buildings and the love affair with their extensive grounds, remembered by some of the first students in the newly sited Edge Hill at Ormskirk. The Ormskirk Advertiser[1], Edge Hill College Magazines, questionnaire responses and personal memories are used to help reconstruct the people and geography through their experiences and memories of the inter war period.

The New Building

In 1931 Alderman J Travis-Clegg, the Chairman of Lancashire Education Committee, laid the **foundation stone** of Edge Hill in its present location in the College entrance. The senior student, seconding the Principal, Miss Smith's vote of thanks, said 'We build in ideals as well as stone, and I hope that the spirit of the old College, which has bound in unity successive generations of students, will continue to flourish within these newer walls[2]'. Ron Tootle[3] has in his possession a photograph of his father working on the foundations of the main buildings.

Perusal of the *Ormskirk Advertiser* from 1931 to 1939 shows a dominantly agricultural flavour; sugar beet, and potato crops are regularly referred to, and the paper had a regular 'Poultry Notes' feature. This latter point struck a chord with J B Priestley, who in his 1933 book, *An English Journey*, commented that 'the whole of Lancashire appeared to be keeping poultry'.[4] There are frequent mentions in the newspaper of the National Farmers Union (NFU) with concern expressed over fowl pox vaccine, and Colorado beetle, a severe pest of potato crops. In 1931 a new Morris Minor car was advertised for £100. In 1933 the (pea) canning industry and Bickerstaffe colliery are mentioned, Gracie Fields visited Ormskirk,

---

[1] It is interesting to note that the full title of the Ormskirk Advertiser includes Southport Advertiser and Agriculture and Mining Intelligencer for West Lancashire
[2] Edge Hill College Magazine 1954, No 63, in celebration of 21 years at Ormskirk
[3] Personal communication 30 August 2004
[4] In Chapter 8, Lancashire, Priestley concentrates on towns. He started his journey in Liverpool, then Manchester and Bolton, to Blackpool, and back to Preston and Blackburn. He was very disparaging about the Lancashire cotton industry, 'a mere upstart to wool'

pickpockets were operating on the market, and farmers were urged to keep more pigs. Ormskirk was a place where people knew each other; Lil Moulson remembers farmers meeting to discuss business in the back room of the Plough Inn on market days.

On 2 October 1933, the new buildings, designed by Stephen Wilkinson, Lancashire County Architect[5], the main H-shaped block, incorporating four halls of residence and dining hall, and a separate indoor swimming pool, were opened. An article in the *Ormskirk Advertiser*[6] headed 'The new Edge Hill Training College, to be opened by Lord Irwin on Monday next', delivered a 'full description of the imposing building'. The old buildings were said to have long been inadequate and cramped, with no recreational ground or space for development. It was described as a red-letter day that 'Ormskirk has now become an important centre for higher education'. Lancashire authorities were said to have chosen a convenient approach by bus and train.

> '...delightful site... (on) the breezy lower slopes of Scarth Hill, overlooking an unspoiled countryside, sufficiently secluded for studious pursuits and yet within a short distance of the town. A stretch of the green fields sloping from Ruff Lane to St Helens Road was chosen as the site of the college, and here during the past two years the new buildings have been slowly rising until today they stand as one of the finest and most modern of their type, an architectural achievement... Even the lay visitor cannot fail to be impressed by the dignity and beauty of the new building and its open aspect... The College is built of brick with mellow stone facing and is in modernised Georgian style with a classical entrance... having dignified pilaster facings and balconied windows enhancing its stately appearance. The entrance... is by... handsome iron gates. The Porter's Lodge, a neat and substantial brick structure stands in a nicely laid plot within the gates and beyond it a long straight drive, bordered with poplar trees,[7] leads to the College...
>
> Handsome oak doors... give access to the interior. This central block contains classrooms, laboratories, handwork rooms, bookstore and staff room. The left hand portion has

---

[5] Pevsner 1969:184
[6] Ormskirk Advertiser 28 September 1933
[7] Reports suggest these were actually lime trees, though poplars may have been planted. Certainly the trees did not do very well and were later replaced by John Downie crab apple trees, though these too have been removed

the Principal's room, the secretary's office, and a handsome library furnished in oak, with ample shelving and recesses. In both portions of the block the rooms open out of a spacious corridor walled with biscuit tinted tiles, with borders of Wedgewood Blue. The classroom block forms one side of the quadrangle, and the other sides are formed by the assembly hall on the left, the gymnasium on the right, and the dining hall running parallel with the classroom blocks. Over the dining room are the Principal's flat, the Matron's room, and the sick ward. The assembly hall... will be a splendid setting for college gatherings. It has an oak panelled stage and rostrum, with decorated plaster work in the form of a classical mask. The ceiling is artistically panelled with electric lighting from hexagonal green tinted glass...

The two wings are the residential quarters for students and academic staff. Each wing contains two halls... Each Hall has its own entrance hall and front door, each has also a delightful common room with rustic brick fireplaces and inklenooks... There are cosy furnishings and all the four halls have their own individual colour schemes. Every student has a charming little room, cream walled, with brown woodwork and furnished in light oak. Curtains and bedspreads in green, blue or rose-pink give gay touches of colour... Each hall will accommodate 50 students.

The account goes on to describe the picturesque nature of the oblong lily pond and its cherub sculpture, and notes that there are one hundred and seventy-six students, thirteen staff, matron, secretary, two lady cooks and 'necessary domestic staff'[8]. College societies were listed, as were facilities in the grounds for four hard courses for tennis, hockey and cricket pitches. From the College archives, an advertisement for Furse Electrical Installations, a Manchester based company which provided the lighting, electrical cooking, fire alarm and stage equipment, described the building as 'considered the finest example of its kind in the British Isles'.

Edge Hill College Magazine, 1934: No 43 gives more information about the opening of the new building, as students lined the drive, and Lord Irwin planted a sweet chestnut tree. With a gold key he unlocked the education door, and declared the building open.

---

[8] These staff included four resident maids for each Hall, according to Miss Margaret Bain, later Edge Hill Principal, writing in Edge Hill College Magazine 1955, remembering twenty-one years at Ormskirk, when the 'fine new building was freshly equipped and furnished'

In his speech he wondered, perhaps inappropriately on opening a new building, 'whether now-a-days we give sufficient weight in our philosophy to the continuance of inherited things'. The Head Student reported that 'naturally we will miss many things... such as proximity to Liverpool, with its bustle, trams and shops, but in Ormskirk there is much to compensate for that loss'. The new building was partially finished on September 23, when supper was followed by dancing in the gymnasium. On October 2 the building was floodlit, and 'Advanced History students gave accounts of their findings about the great people whose names distinguished the four halls'. The halls were named Stanley, Clough, Lady Margaret and John Dalton, in honour of the Derby family and after three individuals famous in the history of Lancashire and Education. Stanley was the family name of the Derby family, and the first Earl's wife was Lady Margaret Beaufort, John Dalton was a famous Manchester scientist, while Ann Jemima Clough was a pioneer of higher education for women, having founded Newnham College, Cambridge (Montgomery 1997:47).

Fig 3.1 The H shaped Edge Hill buildings: 1935 air photo
Source: Edge Hill archive

The 1930s College buildings have Carboniferous sandstone facings, pillars and crest, and roofs of Westmorland green slate. The Archway of the Education Building, 'supported by four Corinthian pillars with beautifully decorated capitals is surmounted by very fine carvings of the Lancashire Coat of Arms and the motto *In Consilio Consilium*[9]. Internally the furniture was solid light oak, designed by Waring and Gillow of Lancaster, and the cutlery was silver-plated. The new buildings are shown on an early aerial photograph housed in the Security Office in College, and taken with a plate camera from a small plane in 1935, probably reflecting Council pride (Fig 3.1) It shows the buildings standing in open space with newly planted trees along St Helens Road, and the scene is strangely devoid of either people or traffic.

The Glorious Grounds

The big advantages of the new site were its modern buildings and proper playing fields. From the confines of Liverpool, both staff and students of Edge Hill rejoiced in the open space on the Ormskirk site: it was almost a love affair. All contributed to planning the conversion of thirty-four acres of farmland, adorned previously by a few hedges and three trees, into an attractively laid out and productive flower, fruit and vegetable garden. There was a rose garden[10], shrubbery, orchard, flower borders, seats, sundial and summerhouse, with beehives and poultry. Mr Steer, County Horticultural Adviser, visited and spoke about how natural features in the gardens should not be obscured, while a farmer from the back of the College helped with the construction of the 'wild' part of the grounds. Miss Smith liked the new site, and walked round daily;[11] she donated an orchard on her retirement. Many gifts of bulbs and trees were donated and both staff and students appeared to indulge in a virtual orgy of planting. Great pleasure was derived in particular from planting unusual trees. Morreys of Kelsall, sited like Edge Hill on New Red Sandstone, on the mid Cheshire ridge, was the nursery that provided the original plantings. The kitchen garden kept the College self sufficient in vegetables, and the grounds were used for gardening classes. Flowery language, in the style of a Victorian Country Diary, in successive Edge Hill College Magazines expressed 'never-ceasing delight' in the conversion of farmland into productive gardens and restful spaces, noting the colours of flowers in the borders and the attacks of the

---

[9] Edge Hill College Magazine 1935, No 44
[10] In September 2004 during work on the new Education Block beyond the lake the Rose Garden was converted into a car park. The genetic stock of the old specimen roses has been preserved and a new garden is proposed
[11] Edge Hill College Magazine 1961, No 70, reported the death in 1961 of Miss Smith, Principal from 1920-1941

local hare population on their plantings. From 1934 to 1938 staff must have become much attached to the lovely peaceful and green surroundings of the College site in Ormskirk, as the following extracts of 'In the Grounds' reports show. The accounts of the garden where staff could sit in peace and admire the view provide a contrast to the grounds where students worked to provide flowers, fruit and vegetables. The Edge Hill College Magazines of these years represent a resource in what is now called phenology, that is recording the dates of the seasons, to check if the climate is changing. For example 1935 was described as a difficult year, 1936 a hard winter, with cold and sunless spring and late summer, while there was a drought in 1938 when water pipes were installed. There are also records of nesting birds, swallows in a hall doorway, and moorhens rearing young in the lily pond. Even the quarries at Parbold from which the rockery stone was derived were noted.

> **1934:** The spacious grounds and wide horizons …are a never-ceasing delight. Two years ago our thirty four acres were open fields where rabbits scampered. A year ago they were a builders' dump of sand, brick-heaps and mortar. Today they offer us… the green restfulness of lawns and playing fields… we have gathered our gooseberries… and radishes, and potatoes and spinach… The Principal has given us an orchard: beautiful trees… - a mulberry, medlars, quinces, walnuts, Japanese cherries, apples and pears. Mrs Booth has given us a plantation of conifers and some unusual trees… And Miss Chadwick has given us a tulip tree… and Miss Le Fevre a prunus tree… there is a book in which even the smallest of our garden gifts is recorded.

> **1935:** …this has been a wonderful year in the grounds. We wondered about the stone: and when it arrived we saw that it was beautiful - local sandstone from Howitt Quarries at Parbold - of many shades of brown and yellow with warmer streaks of pink and red. And it was not in slabs and geometrical shapes but in huge and irregular blocks, with the grains glittering in the rough surfaces… We saw the summer house being erected… There have been plenty of vicissitudes in this garden making. Besides drought, gales, late frosts, birds and weeds... there are …hares. We see them loping across the grass a few yards away in the mornings and moonlit evenings and we find they have eaten all the shoots of carnations and pinks, even through wire netting…

> **1936:** The Rock Garden is now complete… The hollow is now enclosed by its semi-circular sunken wall of Parbold

stone... A beautiful type of juniper tree grows beside one set of steps... The three (bee) hives have been full this year and we hope they will truly give us 'honey for tea'... A weeping ash has been planted... A sundial on an old stone pedestal has been placed where the broad paths meet. ...An ancient Essex Manor House provided the pedestal... in the shape of a carved chimney pot of apricot coloured stone. Local sandstone provided the base for this pillar and though the old copper dial bears the name of a London maker it had marked the hours in an old Lancashire garden for a century or more before it came to Edge Hill... A piece of ground near the St Helens road entrance to the kitchen garden was given over to Mr Steer for his classes for Teachers which met every Saturday morning in College and in the summer term on Monday evenings too... The hares have multiplied, the rabbits have burrowed...

1937: With the coming of spring there was colour everywhere ... From earliest spring the rockery has been a mass of blooms... several small gifts of interesting rock plants have made their appearance recently... On the far side of the rockery the Students' Avenue is now completed and a lovely vista along a grass pathway between flowering tress leads to the Kitchen Garden beyond... the Biology Materials class did valiant service... planting lavender cuttings and young cabbages... The great constructive feature of the year has been the formal rose garden... A rectangular lily pond has been made... Six new garden seats have been bought... In the Kitchen Garden there has been a bumper crop of gooseberries, raspberries and currants... For the Staff Garden more plants have been received...

1938: The household has been supplied with green vegetables right up to the middle of June. Water has been taken up the garden in pipes... saving many crops... Four apple trees given by the students were planted in March and form the beginning of a line of apple trees that will ultimately form a leafy walk by the beech hedge at the top of the hockey field. Flowers for the dining room tables, for the Halls and Students' rooms have been available all the year... The School Garden supplied a 'posy' of mixed flowers for the Principal at the Students' Garden party... In the rose garden the climbers have made good progress towards covering the pergolas... and the paths... have been completed with a covering of Ruabon brick chippings of an attractive red colour which contrast well with the vegetation. Opposite Clough Hall a border of violas has been made along the lawn outside the gymnasium... The cold summer has prevented much outdoor study …but the summer houses have afforded welcome shelter when it was possible to sit outside.

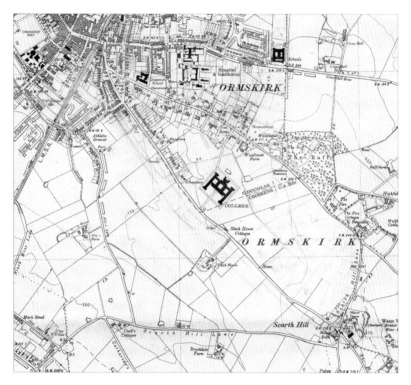

Fig 3.2 The new College buildings. Reproduced from the 1938 OS Provisional Edition Lancashire sheet XCII NW six inches = one mile

Source: NGAS Map Collection

The new Edge Hill College buildings are shown on a 1938 OS Provisional Edition map in the comprehensive geography map collection (Fig 3.2). The map clearly shows the 'H' structure of the 1930s main block, with separate porter's lodge on St Helens Road and swimming pool, with the Ruff Lane exit between the houses of Levens and San Remo. Footpaths across the former fields are still shown, though Mount Pleasant appears missing. The map shows 'CA' boundaries, probably river catchment areas. A straight line across the campus shows the College in the Douglas catchment, with Crossens to the south.

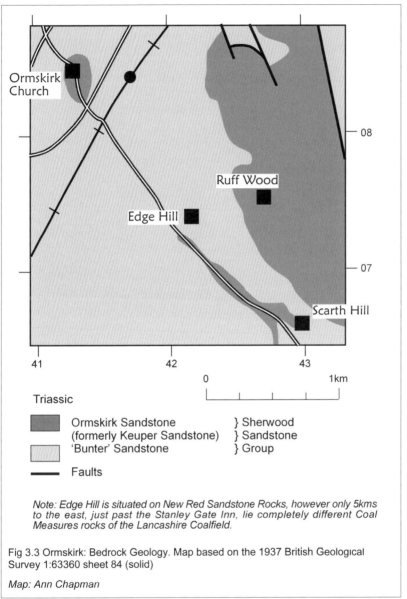

Triassic

Ormskirk Sandstone
(formerly Keuper Sandstone)   } Sherwood
'Bunter' Sandstone   } Sandstone
     } Group

Faults

Note: Edge Hill is situated on New Red Sandstone Rocks, however only 5kms to the east, just past the Stanley Gate Inn, lie completely different Coal Measures rocks of the Lancashire Coalfield.

Fig 3.3 Ormskirk: Bedrock Geology. Map based on the 1937 British Geological Survey 1:63360 sheet 84 (solid)

Map: Ann Chapman

Hidden beneath the Ormskirk site is the **bedrock**, an important sandstone aquifer, with naturally filtered water held in the pore spaces between its constituent sand grains. A bedrock geology map shows that Edge Hill College is sited on the soft Triassic New Red Sandstone, the 'Bunter Sandstone' of

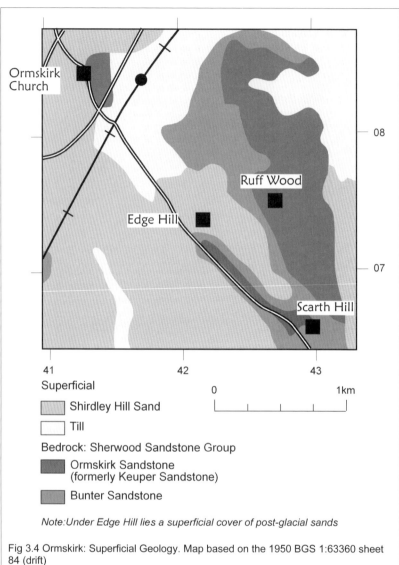

Superficial

▨ Shirdley Hill Sand

☐ Till

Bedrock: Sherwood Sandstone Group

▩ Ormskirk Sandstone
(formerly Keuper Sandstone)

▩ Bunter Sandstone

Note:Under Edge Hill lies a superficial cover of post-glacial sands

Fig 3.4 Ormskirk: Superficial Geology. Map based on the 1950 BGS 1:63360 sheet 84 (drift)

Map: Ann Chapman

the Sherwood Sandstone Group (Fig 3.3) though no solid rock outcrops on the College site, as it is everywhere covered by superficial Shirdley Hill Sands (Fig 3.4) which covers much of the Ormskirk area. Small outcrops of the overlying Ormskirk Sandstone (formerly Keuper Sandstone) occur under Ormskirk church, in the quarry at Ruff Wood and along St Helens Road from College to the slight ridge of Scarth Hill, where an exposure of cross bedding in the sandstone is recognised as a Regionally Important Geological Site (RIGS). Small outcrops of Ormskirk sandstone emerge from the remaining boulder clay cover shown on the superficial geology map; discerning eyes may detect these bedrock outcrops in the slightly higher land along the St Helens Road, and in the slight ridge of Scarth Hill. The outcrop here is a significant field visit site.

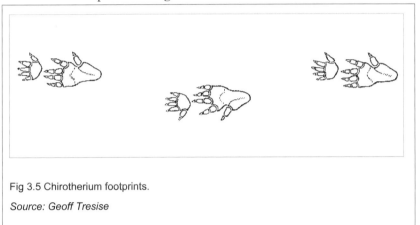

Fig 3.5 Chirotherium footprints.

Source: Geoff Tresise

Chirotherium – the invisible dinosaur

In the 2003 excavations for the lake on Edge Hill's 'Western Campus' a cross section of the underlying Triassic rocks was obtained. Brightly coloured bedded rocks were exposed; layers of bright yellow, white and deep red sandstones stained by iron oxides, which had been formed in a tropical desert environment. Among the dominantly sandy layers was a thin layer of finer mudstone, indicating that in the past, about 200 million years ago, the area had been a temporary lake, at a time when dinosaurs such as Chirotherium walked and in some places left their footprints in the soft mud at the lake edge. Under the hot sun, the mud dried and was later covered with sand preserving the footprints (Fig 3.5)

Chirotherium was a primitive dinosaur known only from its hand like footprints (chirotherium means hand animal) discovered in a building stone quarry at Storeton on the Wirral in 1838 (see Tresise 1989 for the full story) Reconstructed from the footprints, Chirotherium was an early ancestor of crocodiles. The rocks at Storeton are exactly the same age as the Ormskirk Sandstone underlying the College site, but unfortunately no footprints were discovered before the lining for the Edge Hill lake was backfilled. Geography staff did however feature in the non-news in *The Ormskirk Advertiser* of 'No dinosaur footprints found'!

Fig 3.6 The old Scarth Hill water tower.

Source: Mona Duggan

The Scarth Hill **water tower** on the sandstone outcrop was a prominent Ormskirk landmark, and a popular focus for Sunday afternoon walks both for Edge Hill students and people in Ormskirk (Fig 3.6) In the 1930s it was a quiet and safe walk around the campus and through Ruff Wood to the water tower. The 1879 water tower, known locally as the Pepperpot, was attractively and elegantly proportioned: Pevsner 1969:184 considered it the finest building in Ormskirk, with its very tall arches to support the water tank. It was circular in plan with eight circumferential brick pillars surrounding a central shaft. Its function was to store treated mains water in an elevated tank, to give a 'head' of water, so that water would flow out of taps when they were turned on. The modern concrete 1960s 'wineglass' tower, typical of rural areas, is variously described as brutal and regrettable, but the old tower, built by four suppliers, could not be amalgamated for modern water supply.

Learning in a New Place

Writing in 1965 Tom Eason commented that at the new site in Ormskirk Edge Hill's curriculum was narrower, the College perhaps more inward looking and the vision more sharply restricted to elementary education[12]. Times were certainly not easy in the 1930s depression years. It was difficult for 1935 leavers to get jobs, and Edge Hill's numbers were reduced; yet in 1937 there were still forty-

---

[12] *A History of Edge Hill*, the first eighty years, in Edge Hill College Magazine 1965

one vacant places (Montgomery 1977:) Paper was scarce and teachers drew maps on the blackboard. Geography was still integrated (human and physical) with a focus on landscape, regional studies and land use. Textbooks were in short supply, and in those days went through many editions. L Dudley Stamp was a prolific and influential writer of geography textbooks, on *Commercial Geography, Glossary of Geographical Terms, Regional Geography, The Land of Britain: its Uses and Misuses*, and *The British Isles*. His 1929 regional geography book *The World* had an amazing longevity with an eighteenth edition in 1966. Stamp also directed the 1930s Land Utilisation Surveys, using schoolchildren to survey their local environments as an exercise in citizenship and based on a belief that geographers neglected the UK countryside.

The Use of Broadcast Lessons (after Going 1938)

Against	*In favour*
Children's attention not held	Local colour & background supplied
One way track	Broadened outlook, stimulated interest
Only a fleeting impression created	Brought in touch with realities
No illustrations	First hand information supplied
Difficulties over time, reception and fitting into school timetable	Children influence by the knowledge that thousands of others were also listening

The accent in textbooks was still on the value of a knowledge base in geography, though there were trends towards more understanding of causes and effects (Walford 1996). The few textbooks in schools in the 1930s were partly compensated by new resources for teaching. In November 1933 Mr Dunkerly of the BBC had visited Edge Hill to demonstrate a 'wireless' lesson, on the life of Buddha, while by 1938 over 9,000 schools listened, to BBC **Radio broadcasts**[13], Travel Talks and Regional Geography, though geographers hotly disputed their value. An article in *Geography* in 1938 pondered the effect of the 'wireless' on the national character whether radio broadcasts provided 'a field for appreciation and illumination' or led to the 'acquisition of a mass mind'. Concern was expressed over the encouragement of 'intellectual passivity' by listening to the radio, as opposed to whether it encouraged new interests. There was

---

[13] I can remember a radio being brought out in a primary school lesson, possibly geography, in the mid 1950s. The radio was flat 2-foot square brown box with a central speaker, which was put on a desk for us to look at. I think we had to listen to the broadcast in still silence, then write down what we could remember afterwards (SW)

conflicting opinion as to whether human or physical geography should be stressed in radio programmes, while questionnaires by the BBC and teachers sought answers to these questions. There appeared to be no clear answer to the vexed question of whether pupils should take notes or not during radio programmes, though a variety of follow up activities were advocated. The Travel Talks did not suit all teachers, and the Regional Studies[14], for older children, were not very popular. Going 1938 comments that perhaps the 'talking film' will shortly be a commercial proposition, and the 'wireless talk without pictures[15] will be like an ordinary lesson without the blackboard'. Pros and cons of radio lessons were summarised as shown in the table.

Geography at Ormskirk

The front of the main College building housed the Education block, in which **geography** was taught. 'The Geography Room, 'in which as the Principal had amusingly announced, you could see your past' was always full of merry groups of students viewing again photographs of themselves and others' (Fig 3.7). Geography was optional at this time. Patricia Ball (née Fearenside 1936-1938) had to give up the subject when she took Matriculation. **Eileen Boocock**, a student from 1935-1937 remembers being taught geography in the Education Block (Fig. 3.7) by Miss Butterworth who was later to become Principal of the College after Miss Smith, recounts her story next.

Black Stockings and a Black Hat

Eileen Boocock 1935-1937

When asked what did we learn in geography at Edge Hill, it is easier to answer - what we didn't study! It was a very comprehensive course. We covered the whole world - continents, climates, resources, peoples, map making, map reading and symbols etc. I remember fieldwork to Derwentwater in the Lake District, short local visits to Parbold and my own explorations on the River Lune from source to mouth, as I had read about the river changing its course. We looked at mountains, and did quite a lot of hiking, following contour lines on

---

[14] Going 1938 lists the Radio Talks given in 1935-36. Regional Geography featured 'Monsoon Lands', India, China, Japan and East Indies, while Travel Talks were based on European locations
[15] I think I remember teaching geography lessons in the 1970s attempting to co-ordinate radio broadcasts with showing slides, or even possibly a filmstrip. The subjects were either Third World or European topics

maps. Much of the geography was done in a special study convenient to our homes, much as a modern dissertation. I found Miss Butterworth, the only Geography teacher at the time, a charming lady, very fair and straight. On the geography trip to the Lake District Miss Butterworth allowed a diversion past Carnforth and Morecambe so that I could wave at my home in Heysham. I remember one geography class when Miss Butterworth made each student speak from the front. My topic was lentils, which caused some giggling, as we ate a lot of lentils.

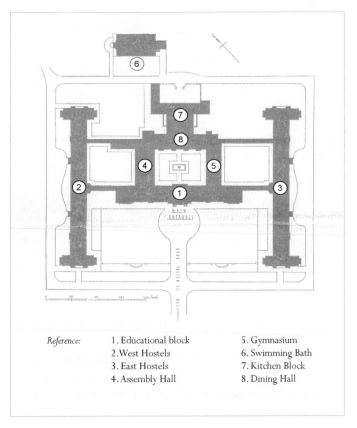

Reference:	1. Educational block	5. Gymnasium
	2. West Hostels	6. Swimming Bath
	3. East Hostels	7. Kitchen Block
	4. Assembly Hall	8. Dining Hall

Fig 3.7 Edge Hill's main block: Geography was taught in area 1

Source: Edge Hill archives

I thought Edge Hill College very smart and elegant. I did Advanced Art and Gardening, though all students had to volunteer to do extra gardening. The College had greenhouses and hens, and Miss Butterworth kept bees. I have vivid memories of a watercolour I did. We walked past fields all the way to Skelmersdale (no New Town then) to paint in a very beautiful dell. Life was very Spartan. The water was very cold in the swimming pool when we did lifesaving. I never regretted going to Edge Hill, but it was hard for people who were not tough, though the results were good. Miss Smith was a dragon, very stiff, even with the staff. Students had to be on the high table, rotating to be opposite Miss Smith. Conversation with her was excruciating. I found that Miss Smith was interested in birds, and I used to talk to her about oystercatchers, knowing about them from living in Heysham. The other students were grateful they didn't have to listen to Miss Smith's remarks.

I remember going to the Picton Hall in Liverpool for some evening lectures, and school practice was in Liverpool, with observation in Ormskirk. On school practice we were required to wear black stockings and a black hat, without which we were considered unable to get a job. I made myself a black (astrakhan) fur hat, but Miss Smith disapproved and made a speech to the whole College about it. When I got to one school practice in Maghull or Waterloo, the children were all sitting on piles of atlases as they had no seats. One sad instance was a poor little boy, who looked like a waif and stray, and his mother had sown him into a vest made of newspapers to keep him warm over the winter. I was horrified that people were so poor they had to do that.

When we left Edge Hill we were given a book as a souvenir with all the names of the students and staff. There were 54 children in my first class. The previous teacher had had a nervous breakdown as she couldn't control them. After that anything was good. I realised how much things have changed when I was talking to some Lancaster University students on the train. They were complaining about how far the University is from the town. I replied that there is a very good bus service, but they said no, they had to have a car. I said that we walked into Ormskirk and bought egg and chips for half a crown, in a cafe which had a fire, as Edge Hill was always very cold.

Like Eileen Boocock, Dorothy Fox (née Elliott 1938-1940)[16] remembers the rising bell at 7.00, morning lectures from 8.30am to 12.30pm, and evening lectures from 4.30 to 7.30pm with a quiet hour after supper and lights out at 10.30. There were lectures on Saturday morning and compulsory games each day and swimming on Monday evenings. Alice Humphries remembers that 'the Edge Hill days were

---

[16] Letter 6 October 2004

tough[17]. She remembers the view from Stanley Hall windows was 'rows of cabbages and the water tower. Windows in Lady Margaret Hall looked over the newly planted rock gardens and gooseberry bushes. Students had to pick and 'top and tail' the fruit'[18]. From Stanley Hall today, the view past the water tower shows a clear rise to the sandstone outcrop on Scarth Hill, though the view in the opposite direction, from Lady Margaret Hall, is now obscured by trees and new buildings.

The annual Geography all day 'expedition' to the Lake District appears to have been eagerly anticipated and must have been a welcome break from the normal work routine; certainly their descriptions in Edge Hill College Magazines of 1934 to 1939 were extremely long, giving full details of the weather, the route taken and the geographical changes noted along the way, noting also the route walked, and where picnics and tea were taken (Fig 3.8). Typically coaches would set off at 8.30am and return at 9.30pm. In 1934 geography students visited Langdale, in 1935 from Ambleside they walked to Grasmere and Rydal, in 1936 from near Keswick they visited the stone circle and walked up Castle Head Hill, which gave a good view over Derwentwater and Bassenthwaite lake, while in 1937 students climbed Wansfell in heavy rain. The precise destination varied each year, as though Miss Butterworth was using the visits to explore the Lake District while by the last pre war visit in 1939, she left the students 'to examine Stock Ghyll Falls, and other phenomena nearer the town' (Ambleside) while the students walked round Loughrigg. These expeditions were a local regional geography type of activity, looking at rocks and glaciation, with a 'competition' to see who could gather most evidence of human habitation. They were fairly typical of the time when the countryside was seen as an area for geographical study and fieldwork was linked to nature study and admiration of grand scenery (Marsden 1976). Only in 1938 was there a different destination, the SW Lancashire coast, where at Hightown the submerged forest was seen. The following gives a taste of the nature of these rather effusive accounts.

Geography Expeditions

1934: The Geography Expedition to the Great Langdale Valley on June 29th was both a social and educational success... The party travelled by coach, and the main route

---

[17] Edge Hill Newsletter 2002

north was taken, via Preston, Lancaster, Kendal, Windermere and Skelwith Bridge to Langdale and the Great Langdale Valley. Here the party... walked... to Blea Tarn... The journey, besides being in itself delightful, had afforded many opportunities for the practical observation of geographical principles as evidenced in the scenery, wild vegetation, agricultural activities, and the building materials used in relation to the local rocks. Many evidences of former glaciation were seen... The delightful change of scenery from the flat plain to the low fells and then to the wild rugged pikes and crags surrounding the Langdale valley was much noticed... Some people visited Dungeon Ghyll and some were content with Mill Gill, which, owing to the wet weather a few days previously, were a little fuller than in this dry summer had been expected. The return journey on a mild summer evening as the sun was just setting was a delightful climax to a lovely day.

1935: ...an interesting competition was held to see who could find most evidences of human life amongst the grandeur of nature... The successful competitors had twenty five different evidences... including stone walls, a boat on the lake and a wireless pole.

1936: ...The party set out in two coaches... and... stopped for a time at Ambleside and while having lunch in a meadow, others went down to the lakeside and had a look at some of the Roman 'Remains' there. We then left for Keswick noticing how the smooth hill contours around Windermere gave way to the rugged outlines of the volcanic rocks between Ambleside and Keswick. We drove over Dunmail Raise... (but) ... we left the coaches in order to do some walking... We reached a Druid Circle and having examined that and observed the things of interest in the landscape we descended to Keswick. From the outskirts of the town we climbed Castle Head and although the climb had been stiff, we found it well worthwhile. We had a magnificent view over Derwentwater and Bassenthwaite lakes and the town of Keswick... we had seen six lakes in one excursion as well as three types of Lakeland landscapes...

Fig 3.8 1934-1939 Geography expeditions to the Lake District.

Source: Sylvia Woodhead

1937: ...we started out... well armed with maps, papers and provisions, and although the weather was dull it did not damp our enthusiasm. But the long history of fortunate weather for this expedition was broken... Stock Gill Falls were seen and photographed and the party then enjoyed the scramble up Wansfell Pike in spite of driving rain and accumulating mist... We made our way down to Troutbeck, the rain still falling, and so by the valley to Windermere and there we had a much needed tea and were a very cheerful party in spite of our damp things. We rejoined the coach at half past five, changed into dry footwear and drove home in record time... We were all disappointed because the weather prevented us from seeing what we had intended to see, but we gained some knowledge of physical features, rocks and climate, and we certainly enjoyed it.

1938: On... May 30th, a Geography expedition was organised by Miss Butterworth, the object being to examine some of the features of the South-west Lancashire coast, between Hall Road and Hightown. The journey, by coach, to the coast was through flat drained moss-lands... The drainage channels were noticed and the black peaty soil... On the beach at Hall Road the precautionary and defensive measures which had been taken to prevent the sea from encroaching further inland were noticed... we were able to walk along the shore to Hightown... The point of greatest interest to all was the exposed forest bed which had extended along the coast. The glacial deposits of boulder clay have been washed away revealing trunks and roots of trees, parts of the forest beds, which extend far inland under the clays and sands. The trees were wonderfully preserved, the lenticel markings on the bark being easily distinguished... While we had a picnic tea on the dunes we saw an astonishing sight in the sky which we afterwards found was the finest display of a double solar halo that had been seen for years. (On) The return journey we could see the outcrop of Millstone Grit in the neighbourhood of Parbold looming up from the flat surrounding country.

1939: ...A party of forty-six excited Geography students set out on a long-anticipated visit to Ambleside... The first drumlin was the cause of much excitement and the first glimpse of the distant Lake District hills was a thrill even to those who had visited them before. The journey along the peaceful lakeside from Windermere promised great pleasure ...and on arriving at Ambleside we parted from Miss Butterworth, who was examining Stock Ghyll Falls... At the top of the first slope we paused for a moment to look for evidences of an ice age and we were fortunate in seeing a cirque, hanging valleys and the erratics which characterise the district. ...we had a wonderful view of Grasmere ...we took a last look at the peaceful splendour around us …and were soon back in Ambleside where we had tea. (M. Barker).

Something of the character of the Lake District, slightly later than Miss Butterworth's geography expeditions with Edge Hill, may be gained from the book of photographs taken in the 1940s and 1950s, recently published by the Friends of the Lake District. Gwen Bertelsman, a visitor with an Austin Seven car and a camera, took a series of 'outsider images' around the Lake District, which record the wartime Plough Up Campaign, and show hand grass scything, horse drawn ploughs and whole families involved in the oats harvest. People are shown skating on Windermere in the harsh winter of 1942, and picking bluebells on Loughrigg Fell. Her photographs show that there were very few visitors, roads were poorly surfaced and very narrow; most of the traffic was walkers, equipped with hats, coats and stout shoes (Varley 2003).

Marjorie Myatt (Barker) 1938-1940 enjoyed her Advanced Geography classes. She had her own atlas that she shared with Edna Sanderson. She remembers Miss Butterworth, a motherly figure, showing slides of different parts of England, which students had to identify from clues like stone walls. Marjorie still has her thesis on the Potteries, which then were coal fired. Irene Mackenzie shares her experience next.

'It was so strict'

Irene Mackenzie 1936-1938

I was a student at Edge Hill from 1936 to 1938[19] when the Ormskirk site was new, and took Advanced English and History. I only did a little bit of Geography, which was taught by Miss Butterworth, who was a fine looking woman, grey–haired, possibly formerly sandy, but knocking on. In fact all the Edge Hill tutors seemed pretty old to me; I would have liked some younger ones. Some tutors needed pensioning off; they were too old for the job. Miss Smith was Principal; she had such a sour face. She aimed to keep young girls pure, and she was far too strict.

I was formerly Irene Edelston, one of a family of nine, and came from a farm in Goosnargh near Preston. Edge Hill College was a fine looking building, new and modern, a pleasure to go to and to think of it as my College. We were not used to mod cons and I

---

[19] Telephone interview 19 November 2004

enjoyed the room I had to myself in John Dalton Hall, which had electric light and baths. The furniture was light oak, very good quality; we had to look after it. I was quite sporty and played hockey and was in all the teams, but it was not a wonderful time. We worked jolly hard, but made friendships for life. When we went to Ruff Wood to paint, we had to carry our easels, but it was lovely there. The flat landscape was all right, but we didn't get out much except for Teaching Practice. We all had pushbikes, and cycled to Teaching Practice, wearing a black hat. I would cycle from the farm in Goosnargh to College, a distance of about 25 miles. On the afternoons when we were allowed to go out, we went to the pictures in Ormskirk for three pence, and Swift's Café, which was quite nice. I later taught physically handicapped children for five years, those who had had polio, TB or rickets, which were prevalent at the time.

Two stories illustrate how strict it was pre-war at Edge Hill. One girl had got 'in the family way'. It was very much 'hush-hush'. We were all gathered into the Hall, where Miss Smith talked about 'this terrible thing'. The girl had to leave, although she had given an excellent demo lesson to us all that day and had the makings of a gifted teacher. Another girl who was late reporting back in at 10.00pm on Saturday, on investigation proved to have been in a car with a man! What a carry on! It was different then. As I finished at Edge Hill in 1938 the war was looming. It was to change our lives forever; nothing was the same afterwards.

From 1939 during the war, the Home Service broadcast a special news commentary for schools (Plosjazski 1999:62). Some schools received visits from Forces personnel for first hand reports, while others had correspondence and article exchange schemes. In the 1930s the Institute of British Geographers (IBG) was set up to publish, in their Transactions, the research of geographers in universities. The fourth monograph to be published was Dr Alice Garnett's 'Insolation and Relief; their bearing on the human geography of the Alpine regions', whose title suggests a climatic determinism approach. Three of the early Presidential addresses dealt with regional geography (Steele 1960). These values were stressed in the geography taught at Edge Hill before World War II.

> '...we embarked on our new life in the country and began
> to explore its possibilities and meet its difficulties. In our
> happiness with our new environment we could not then

have believed that we should have to move again in six years'.[20]

The Campus at War: 1939-1945

After the long report on its opening, Edge Hill College does not appear to feature in the *Ormskirk Advertiser*, which in 1938 carried an article on slum clearance in Ormskirk, and in 1939 had advice for potato growers. On 7 September 1939 war broke out, and the much-reduced *Advertiser* detailed blackout regulations, requirements for Lancashire to cultivate 40,000 more acres, fuel rationing, economy measures and air raid shelters. There was no mention in the newspaper of evacuation of Edge Hill (staff and students had moved to Bingley) or its use as a military hospital.

The fate of the Edge Hill buildings during the war has been difficult to trace. It became a Military hospital and the geography building was constructed; yet details are elusive. Some glimpses of Ormskirk at war may be gleaned from the **Minutes of Ormskirk Army Social Welfare Committee**[21], which was set up in April 1940, with the initial remit to provide some 'comforts' for the troops in the area, such as for canteens and games to prevent boredom, cigarettes and bedroom slippers for patients in the military hospital. Other records say little; the Ormskirk **Medical Officer of Health** Reports noted overcrowding in 1940 from reception of children evacuated from Liverpool, but make no mention of military hospitals. Similarly the **UDC Minutes** for the wartime period make no reference to the Military hospital, noting only a food economy campaign in 1941. Personal memories have been more productive than searches through Army archives, though the Territorial Army Committee records have revealed a few gems. The wartime story includes Peter Stein's investigations on munitions factories and military activities in the local area and Charles Rawding's researches on the plough up campaign in south Lancashire. David Owen recounts his experiences training for the war as a hospital orderly at Edge Hill, and has kindly lent a wonderful long photograph of Army personnel outside Edge Hill in December 1939. Grace Hardisty drove an ambulance to the front door of the military hospital, but went in only very briefly, an experience also recounted by Margaret Goodwill, whose sister Beryl met her husband at one of the Saturday evening dances at the hospital. Dr Betty Underwood, a retired Ormskirk GP, through her

---

[20] Miss E M Smith, former Principal, writing in Edge Hill College Magazine 1955
[21] Ormskirk UDC, Historic Collection (UDOr 15/44) Preston Record Office

researches of the wartime history of Ormskirk County Hospital, has made discoveries about the Military Hospital at Edge Hill: both seem to have served wounded service personnel and POWs. Finally the hospital records receive a comment for those who wish to know the gory details of soldiers' illnesses on returning from the front lines. Peter Stein sets the scene for the Campus at War; his account, of wartime developments not shown on OS maps, reflects the rapidity of the military occupation of Edge Hill's buildings.

The Hidden War

Peter Stein

The image that is held today in most people's minds in the run up to the second world war is Neville Chamberlain flourishing a piece of paper on his return from Germany and professing that 'We have peace in our time'. World war appeared remote and the country could breathe a sigh of relief. One year later, his voice laden with sorrow, Chamberlain announced the commencement of hostilities.

It could be assumed that nothing was happening locally during the years leading up to the war, but in fact, a great deal of planning and preparation was taking place on a number of fronts. After September 1939 things moved quickly, the need for the manufacture of munitions and other equipment became apparent and locations for Royal Ordnance Factories (ROF) were confirmed with Fazakerley, Kirkby, Risley near Warrington among many sites throughout the UK. The design for such undertakings was long established and within a short time they were up and running.

Kirkby Royal Ordnance Factory (ROF)

The factory made shells and bombs, cartridges, fuses, pellets and pyrotechnics, and covered 750 acres and comprised 1000 buildings, eight and a half miles of road, and forty miles of railway track including its own railway station. At its peak 32,000 people worked there, operating twenty-four hour shifts. The workers were mostly women, from as far as Southport, Wigan and St Helens. Thirty lives were lost, while other workers lost their sight or fingers in this dangerous work. (Clark 1995)

If you look at OS 1947 Sheet 100, Kirkby ROF does not exist (Fig 3.9). The tenants of Lord Derby, who owned the land at Kirkby, were quickly brought together in early 1940 and told that they had two weeks to move from their farms as the land was going to be used to build a factory. The site occupies an area of some 2km and was well served by transport as it was recognised that large numbers of people would have to get there from all over the North West on a twenty-four hour basis. The factory remained intact during the war protected by a decoy factory built some two miles away. This dummy factory was actually damaged by bombs on two occasions so it must have looked real from the air.

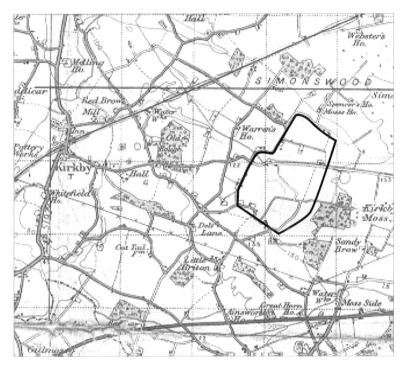

Fig 3.9 Approximate location of Kirkby ROF: extract from OS 1947 one-inch sheet 100

Source: NGAS Map Collection

ROF Kirkby specialised in all forms of munitions filling so storage away from the main manufacturing base was imperative. Two sites at Melling Mount and Moss Farm near Rainford were constructed together with a large Inland Storage Depot at Simonswood. The Melling Mount and the Moss Farm site are still intact and accessible today although the use they were put to in wartime is not recorded. A visit to Melling shows a cluster of buildings set a blast distance apart from each other; this was typical of buildings designed for the storage of bulk explosives. Moss Farm was served by a railway spur from the main Liverpool to Wigan line to a distribution shed. The quality, style and distribution of buildings suggest the site was a packing station where items brought from

Kirkby were assembled and stored for onward distribution. The site was run by a regular Army detachment and housing was available for Officers and Barracks for other ranks. These buildings are currently occupied by local residents.

The Inland Storage Depot at Simonswood was constructed in early 1940 in response to the shortage of storage space mainly due to the amount of bulk material arriving at the docks and the realisation of what damage an air raid could inflict on large concentrations of inflammable stores. The depot was situated in close proximity to ROF Kirkby by the main line from the Bootle docks to Wigan. There was also a spur line into marshalling yards at the back of the ROF. In the late 1980s the line was removed and the rails sold to railway enthusiasts. By the late 1990s in response to increased demand, the complete spur was re-established to service a major fibreboard company.

The ROF at Kirkby was the first of its kind to be earmarked in late 1944 as suitable to be returned to the commercial sector as an Industrial Estate after the war. In March 1946 this was accomplished. This was important as it demonstrated that the government were planning for life after the end of the war.

The first bombs fell on Birkenhead docks on 10 August 1940 and this continued until 10 January 1942. The re-location of businesses and companies of strategic importance was carried out by specially formed committees though there appears to be no longer any record of what was undertaken. From a chance conversation it was found out that what eventually became British Road Services in Bootle was relocated to the old Levens building at the back of Edge Hill College after being bombed out from their Liverpool yard.

There has been a military and aeronautical presence around Ormskirk since Victorian times (Fig 3.9). In Burscough, by the side of the Leeds/Liverpool canal, was a Royal Army Ordnance Corps depot which served all sectors of the military establishment; it is now a housing estate. When the builders moved in, live ammunition and ordnance was uncovered in the soil and had to be made safe before work could continue. Prisoners of War (POW) were also housed in camps around the Ormskirk area. Local knowledge has these POW as being Italian and they worked on farmland throughout the West of Lancashire. There was a camp near Aughton that was designated as a general holding unit for POWs, although little information is available today about its exact location.

Driving from Ormskirk towards Scarisbrick, a disused aerodrome will be noticed. This airfield was commissioned in

December 1942 as HMS Ringtail and was a training unit accommodating all aspects of naval aviation and as a storage area for aircraft disembarking carriers berthed in the Mersey. The station was 'stood down' finally around 1957.

The Meteorological Office scores a minor success story in having a station at Aughton for many years until its closure some eight years ago. Aughton was the first commercial development in the late 1950s in a green belt and was subject to many planning restrictions to prevent it being seen from the main A57 road. However, a tracking radar dish on its roof was something of a give away and became a talking point. The dish was an element of a radio sonde tracking unit and balloon releases were a daily occurrence. The station was part of a chain supplying upper air analyses for aircraft and when stood down some years ago followed a now familiar pattern of becoming – yes you've guessed it – a housing estate. No bombs this time.

The Merseyside of that time was made up of independent councils, Liverpool, Bootle, Litherland, and Kirkby[22] each having their own identity and budgets. Generally they all tried to work together, however this apparent truce did not prevent Bootle Fire Brigade holding on to their machines and refusing to release them to Liverpool during the height of the blitz.

The contribution made to the war effort by Merseyside went on from the first days of the war right through to its conclusion in 1945. Liverpool was considered a prime target for the Germans and documentation and photographs exist from as early as 1935 to substantiate this.

Wartime Edge Hill is brought alive by some personal recollections; Lil Moulson[23], who lived in Ormskirk at the time, was called up in the war at eighteen, and worked at Hattersley's brass foundry, which had been a munitions factory in the First World War. She worked in the office of the foreman, doing timesheets and wages. She lived in rooms above the Plough Inn, which, like the Parish Church, served tea for people during the war, and she remembers MP Commander Stephen King Hall speaking to farmers encouraging the plough up campaign, and also a visit of Harold Wilson to Ormskirk.

---

[22] Kirkby was only a hamlet at the time, with a Parish Council - Lancashire County would have been the decision-making authority. My father, Frank Lawler, was one of the first councillors elected when the Urban District came into being in around 1958 (AC)
[23] Interview (Julia Hedley's mother) 2 Sept 2004

(Territorial Army) Motor Mileage Allowance (1941)

	For first 250 miles in any one calendar month	For mileage over 250 miles in any one calendar month
Motor assisted bicycles	1d	1d
Motor cycles, with or without side car	2¼d	2¼d
Cars up to and including 8hp and tri-cars	4½d	2¼d
Cars over 8hp and up to and including 10hp	5½d	2½
Cars over 10hp	6d	2¾d
The above would not be payable for unlicensed and uninsured cars.		

Although it is known that members of the Territorial Army (TA) were trained at Edge Hill military hospital during the war, perusal of TA archives[24] reveals little about wartime operations on the Ormskirk site, other than early preparations for the war. Minutes of the Medical Services Committee (MSC) on 26 September 1938 reported that Colonel Sandiland had visited the 'Pupil Teachers' Training College' at Ormskirk and had prepared a very exhaustive report, complete with detailed plans, showing how this accommodation could be used for hospital purposes. Apparently 'the buildings in their present condition would accommodate approximately one hundred patients, but further accommodation could easily be added by means of huts'. In the minutes of Ormskirk UDC of March 1940 (UDOR/3/31) is a note that the roofs of the 'Government buildings, the hutments at Edge Hill College, under the supervision of Lieutenant of Burscough Ordnance, were not draining as they should'[25]. By implication it might be assumed that the TA contracted out the building of Block 'A', now geography, and other wartime blocks and huts on campus, in a similar fashion to the military buildings of Kirkby ROF. Betty Underwood, like many others, is intrigued by the H (for hospital as seen from the air) shape

---

[24] Territorial Army and Air Force Association of the County of Lancaster: Proceedings of Committees: search by John Chapman
[25] Email, Betty Underwood 20 May 2005

of the original Edge Hill College buildings, which she had interpreted as part of Emergency Medical Services (EMS) plans, made following World War I, in the Liverpool area for preparations for any future war. Whether there is any truth in this supposition is uncertain.

One of the reasons for finding it difficult to trace Edge Hill's military hospital during the war results from the use of different numbers for the hospital. Nearly every memory calls the hospital by a different number. Peter Stein believes it was the 10th RAMC military hospital; Beryl Reeve called it the 102 hospital, while Lil Moulson remembers Edge Hill as the RAMC hospital, with two divisions, 22nd and 29th. She saw 'walking wounded' patients in Ormskirk in their blue uniform with a red tie and white shirt. Ormskirk UDC Army Social Welfare Committee records refer to the 29th Military Hospital, while the TA Medical Services Committee 14 December 1937 described the hospital at Ormskirk as the 4th General (Military) Hospital, with the 1st at Fazakerly and 6th at Southport. Definitive proof must lie with David Owen who has a photograph of Army personnel outside Edge Hill College, labelled 19th General Hospital, though whether the number did in fact change over the war is not known. Wartime records for the last six months of the hospital simply record it as Medical Division Military Hospital, Ormskirk, with no number. There was considerable secrecy; even in the monthly Army List September 1939, which gives a list of officers on the active list, there is no mention of Ormskirk or its military hospital. However the advertisements in this Army listing, for the Burberry weatherproof coat, Morris Commercial Trucks, a regular service between London and Bombay on the Anchor Line, Norton and Triumph motor cycles, and signs for military roads, give a fascinating glimpse of life at the beginning of the war, while the 1941 TA mileage allowances make for interesting reading, with their mention of pre-decimalisation pennies, halfpennies and farthings.

Plough Up

The outbreak of war resulted in enormous powers being given by Defence Regulations to increase food production in case of a blockade; milk and potatoes were priorities. The Ormskirk area, as recorded by the Land Utilisation Survey 1931-1938, organised by L D Stamp, was noted as being 'one of the greatest potato growing areas of England', Grade 1 Agricultural Land, due to excellent soils, derived from drainage of old mosses, and low rainfall; *brassica* crops (cabbages) lettuces, celery and peas were also considered noteworthy (Fig 3.10).

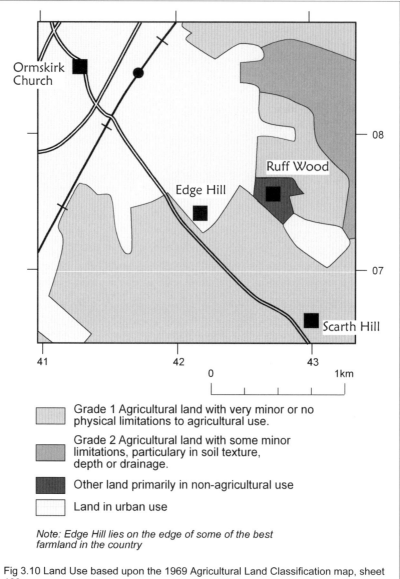

Grade 1 Agricultural land with very minor or no physical limitations to agricultural use.

Grade 2 Agricultural land with some minor limitations, particulary in soil texture, depth or drainage.

Other land primarily in non-agricultural use

Land in urban use

Note: Edge Hill lies on the edge of some of the best farmland in the country

Fig 3.10 Land Use based upon the 1969 Agricultural Land Classification map, sheet 100

Map: Ann Chapman

The first effect of the war was the 'plough up' campaign; farmers received £2 per acre to plough up permanent grassland. Surveys[26] to assist the campaign classified all farms into A, B or C, depending on their percentage of maximum production. Sixty-three per cent of Ormskirk area farms were in category A, eighty per cent of maximum production, reflecting Grade 1 and 2 agricultural land; twenty-two per cent were in category B, sixty-eighty per cent; with only four per cent category C, less than sixty per cent of maximum production. Most category C farms were designated for development and suffered urban fringe problems. Some farms near to industrial targets along the Mersey were badly damaged by bomb craters. Farmers were also told to dig trenches or put up obstacles to prevent enemy aircraft from landing in their fields. Transport problems as the war progressed, resulting in shortage of farmyard manure, made it difficult to maintain high crop yields. Scarcity of labour, with active men away in the forces, was also a problem, solved by school camps, prisoners of war and voluntary land clubs (Rawding 2003).

The extent to which the wartime plough up campaign (Fig 3.11) included the Edge Hill campus is uncertain. Ironically searches at the Public Record Office for the War Agricultural Survey maps and reports are little help, as the map uses the 1927 base map, before the College was built[27], and while there are records for twenty farms in the Ormskirk area, those for the College area are missing. It is recorded simply as 'part grazed' in 1942, suggesting that the College site had not been ploughed up by then, though ploughing orders continued to 1946. It may be that the College grounds were allocated to a local farmer, who may have put animals on the land, although there is no corroborating evidence to support this idea. At Ormskirk County Hospital mental patients were supervised to cultivate potatoes between the wards, but this is unlikely at Edge Hill.[28]

Margaret Bain, former Edge Hill Principal, writing in the 1954 Edge Hill College Magazine, noted that in the war the grounds were neglected, and that the lawns in front of College were ploughed up to grow vegetables. Returning students remember seeing that the hockey fields in front of 'Block A' (now geography) and the main building were ploughed up. It is more likely that the land was leased back to farmers to plough up, than its cultivation being by local people, land

---

[26] The surveys provide a rich data source of individual farms; each has a Schedule number (Rawding 2003)
[27] MAF 73/21/92/1 PRO search by Charles Rawding on 22 November 2004. Edge Hill site is identified as land parcel 373/1, but Schedule records start only at 373/3
[28] Meeting with Betty Underwood, 18 May 2005

girls, or even Prisoners of War: different sources variously describe these as Italian or German. According to Betty Underwood Italian POW were based at Bickerstaffe and had yellow discs on their uniform, while Aughton is believed to have had a German POW camp until 1946 (Rawding 2003). Ormskirk's Land Girls worked on farms, but were excluded from services canteens.

Ormskirk UDC Army Social Welfare Committee provided one night's free accommodation and a car from Liverpool for friends and family visiting the Military Hospital, and arranged lady visitors to the Military Hospital, while the WVS and Rotary Club arranged visits, hire of wireless sets and ran canteens for troops, while the Red Cross and St John's ran a library book service. Other than this, local people were not much involved with the Edge Hill military hospital during the war; they were advised that to talk about it would be against the war effort. 'Careless Talk Costs Lives' and 'Walls Have Ears' were posters of the time. There was great fear of the infiltration of spies, and also of respiratory diseases such as TB, for which the hospital became a centre later in the war. A bomb hit Ormskirk County Hospital in December 1940, making a crater six feet

Fig 3.11 Plough Up campaign poster

Source: Brian Short

by twelve, with no casualties, and Betty Underwood believes that Edge Hill campus was hit during the war and while others have said that the grounds were pockmarked with bomb craters. Six wooden huts were quickly built there to provide accommodation for Service personnel, allies and POWs, in addition to civilian evacuees and air raid casualties, while records reveal cooperation between the two adjacent hospitals. What has emerged is some confusion in local wartime memories between the military hospital on the College site and the nearby Ormskirk General Hospital.[29]

In September 1939, soldiers moved into Edge Hill, and immediately began to convert it into the 19th military hospital of the Royal Army Medical Corps (RAMC). The main building was converted into wards, while the corridors were adapted for hospital trolleys, as were large size lifts. The 1930s Edge Hill buildings form a clear memory for David Owen; they were his home as the 19th RAMC General Hospital from September 1939 to June 1940. Speaking in September 2004, as he recalled his time there at the beginning of the war, a number of parallels with the modern life of students emerged, although discipline was much stronger then. David Owen tells the story of this time.

A Soldier's Story

David Owen

Call up hit me at the beginning of the war, and as I was a psychiatric nurse, I requested the RAMC. After three months training at Aldershot I transferred to Edge Hill. A photograph (Fig 3.13) taken in December 1939 shows me in the left back row, along with other twenty-one year old new recruits, many of who knew little about medical work. In the front row are the Manchester branch of the Territorial Army, and the nurses are the Queen Alexandra's Imperial Nursing Services, with their matron[30], all attired in a grey uniform with red collar trim. Nurses stayed in Stanley Hall.

I shared a room on the first floor of Lady Margaret Hall, with administration occupying the ground floor. When I arrived, the College was already operating as a military hospital, though there were no wounded soldiers at this early stage of the war, and the morgue was not then built. It is quite likely that the roof was painted with a red cross. The wards were in the central block, on both floors, and were for soldiers, all male, who were sick, the nurses were never sick! I was a patient of the hospital myself when I caught German measles. I am positive that Block A (now geography) was not built during my time there. The hospital operated independently of the Ormskirk area,

---

[29] Betty Underwood, 18 May 2005
[30] Miss Cunliffe according to Betty Underwood

except possibly for the provision of supplies, overseen by the Quarter Master, whose stores and kitchens were at the back of the College. The army catering corps did the cooking, while we did our own cleaning; no local people were employed. All College furniture had been removed, and the doors taken off to make the accommodation more like a barracks, and possibly to make room inspection easier. My room had been emptied. There was just a collapsible Army bed, with mattress and pillow, and blankets, but no sheets. I had my kit bag, but no furniture. All windows were covered in black blinds, in order to show no lights. Parts of the front of the College were sand bagged. I started work as a hospital orderly, but we also conducted 'field craft' training for future Army work abroad. Tents were put up in the grounds, and labelled 'casualties' had to be brought in and treated appropriately. I remember my legs buckling on being instructed to give a fireman's lift to a particularly heavy soldier. We paraded in the grounds in front of Lady Margaret Hall, and possibly because the College had a swimming pool, all of us had to learn to swim. I was a non-swimmer, then and now, and I remember being thrown into the deep end of the pool, and having to be dragged out. I still don't like water to this day. On fine days in the spring we did route marches with a small pack, marching 'at ease' round the villages surrounding Ormskirk for fifty minutes at a time, then lying flat on our backs for a ten minute rest. We would set out at 9.00am, lunch was brought out in a wagon, and we returned at 4.00pm, through Ormskirk, now marching to attention

Fig 3.12 Corporal David Owen relaxing in Edge Hill grounds in 1940

Source: David Owen

Fig 3.13 Edge Hill as 19th RAMC General Hospital, 1939 Edge Hill buildings are clearly seen behind the military personnel

Source: David Owen

because we were on display, to sit on the back field for a foot inspection. We did fire-watch (ARP) duty each evening, from the Guard Room at the back of College, with night patrols of the grounds. Every three months there was home leave. I caught the train to Liverpool, ferry to Woodside, and bus to Chester; as we had no vehicles.

Being 'other ranks' I was not allowed to use the front, St Helens Road, entrance. When going to Ormskirk in the evening, for a drink and smoke or to the cinema, the guard on the Ruff Lane entrance would check that all our uniforms were correct. I only began to drink and smoke at Ormskirk, 'just to be sociable'. We would go into a pub in Ormskirk, just for somewhere to go. My friend could play the piano, and as a group of us sang around the piano, local people would buy us drinks. I remember a café at the corner of Ruff Lane, which many of us used. No entertainment was laid on at the hospital in 1939 to 1940; there were no dances. We went for walks, or played football on the grounds opposite Lady Margaret Hall, we made their own entertainment. We normally had to be back by 10.00pm, or 11.59pm with a 'special pass'; I applied to the Regimental Sergeant Major, then the Company Officer would decide whether or not to grant it. I was never late, as there was no reason to stay out later, and we had to be up by 6.00am. There was a piano in the NAAFI, at the back of the College, where alcohol was served, and there was often an evening singsong, although as it was a hospital, we had to be careful not to make too much noise. It annoyed some of us newly commissioned soldiers that we had to salute the nurses, who, as SRN, had the rank of lieutenant. There was no familiarity between soldiers and nurses, and if there was, one or both would be transferred elsewhere. I thought the grounds were really beautiful, and I have a photo of me relaxing in the rock garden near the tennis courts (Fig 3.12).

When David Owen and about fifty others were transferred in June 1940, going first to Warrington before service in North Africa, the plough up campaign had not started. He considers it most likely that the College grounds would be let to local farmers to plough up and that Army personnel would not be involved in cultivating the land. David Owen's account clearly shows the role of the Edge Hill hospital as a training centre, though does not reveal any of the secret duties which Betty Underwood believes may have occurred there later. As it was the beginning of the war there were no actual wartime casualties at that time; patients were soldiers who were called up and became sick. Brian Greaves' father-in-law for example had a hernia operation in an upstairs ward at the front of the College, possibly in 1942[31]. He was sent, in the patients' blue suited uniform, to Scarisbrick for recuperation. Though now over eighty, David Owen's memories are so clear, that he must be correct in his contention that Block A had not been built by mid 1940, although shortly after Ormskirk UDC records note some concern that the Military Hospital huts were not finished and were not gathering water from the roof. However, 'Block A', now the geography building, was built during war time occupation of Edge Hill, probably to a standard Army design, using whatever materials were available, possibly as additional hospital wards. Other blocks were added behind the main building as the officers' mess, with wooden huts (now demolished) constructed as convalescent wards. Wartime buildings did not need planning permission, and few records can be found of Edge Hill's campus during wartime. There is little knowledge of the soldiers or others who lived and worked on the campus during the Second World War. Betty Underwood's researches for Ormskirk County Hospital have revealed that in 1940 fourteen hospital ships disembarked at Liverpool. The first convoy had 194 Polish soldiers from Narvik in Norway, suffering frostbite, who were mostly sent to the County Hospital, and the last one was in 1945. She concurs that transfers of wounded soldiers, many with gunshot wounds, were usually at night, and that there was little enemy action over Ormskirk; three bombs fell, on Edge Hill, the hospital and Hattersleys. In November 1940 the County Hospital had recurrent problems with ambulances and had to liaise with the 29th Military Hospital for transport and in December 1940 had solid wooden gates fitted to the main entrance.

Later in the war, casualties from Atlantic convoys arrived at Liverpool docks, and many were taken to Edge Hill military hospital for urgent treatment before transfer to other hospitals or sanatoria. Towards the end of the war, the military hospital became a centre for treating dysentery and also tuberculosis (TB). Because of the lack of antibiotics, which were not available until the 1950s, both were very difficult to treat. TB in particular was a much-feared killer; those who survived lost much of their lung function, and could breathe so badly they could do little. Treatment involved long periods of bed rest in sanatoria, with fresh and good food: Block A possibly acted as a TB sanatorium.

---

[31] Discussion 3 September 2004

By July 1940 Ormskirk Army Social Welfare Committee had set up a **rota of cars** for concert parties and wireless sets were hired. The British Red Cross and St John provided a library for the 29th Military Hospital, and Mrs G A Hunter and Miss H A Fair were identified as the first librarians. On 6 August 1940 a half fare voucher was obtained for a lady in Ormskirk, whose son was seriously ill at the Military Hospital, while Rev Dr McMillan, Padre at the Military Hospital had written to the Daily Dispatch War Relief Fund direct for cricket equipment, but the request was turned back. By 3 September 1940 a rota of lady visitors to the 29th General (Military) Hospital had been arranged, and a request for £5 for book trolleys for the library resulted in a £3 funding. By 5 November 1940 it was noted that there were nearly 400 patients at the 19th hospital. Visitors brought soap and cigarettes, and there was a suggestion that 'tuition in rug making, knitting and embroidery, handicraft etc will help the men to occupy their time', though it was reported that visiting was not encouraged by the Military Hospitals, 'where rules are necessarily strict'. Committee reports include 'I find that the 29th General Hospital is very short of games such as chess, dominoes, draughts etc', and that 'until matron has been here a little longer, it will be difficult to know how best we can help her'. The Liaison Officer reported he had visited the hospitals 'fairly regularly, but that the Military Hospital organised its own entertainment and has a personal E.N.S.A. producer'.

In March **1941**, two **parties** had been held at Skelmersdale for patients from the 29th, and the WVS had been helping making **black out curtains**, though this latter seems quite late on in the war. On 6 May 1941 the Liaison Officer had been able to arrange a number of **billets** for people visiting the Military Hospital, but it was 'getting a difficult problem, the Town being so full of people', while references were made to **food rationing**. On 3 June, after the Liverpool blitz, **civilian clothing** for the use of Civilian Air Raid Casualties at the 29th General Hospital had been obtained, while the matron and staff of the 29th had 'rendered useful service' at the EMS when the convoy of Air Raid casualties arrived. In July 1941 a parcel of razor blades, toilet soaps etc had been received from Skelmersdale. There was mention that the future evacuation of 26 children and 15 teachers from the area was not the remit of this committee. In August 1941 dances (probably not at Edge Hill) were to be restricted to members of the Forces and Lady Scarisbrick had arranged billets for relatives visiting patients and had arranged midday lunches at the Derby Restaurant, which had proved very successful. She had visited 'five Canadian and two Free French patients, all of whom seemed quite contented'. By

December 1941 discussion included 'woollen comforts', dances were very popular and attended by 'both sexes', though Sunday concerts in Ormskirk were poorly attended.

It is tantalising that for most of the war only occasional glimpses of the operation of the military hospital have been gleaned. It would have been rewarding to contact a nurse or a doctor who worked within the hospital later in the war. What is intriguing is why Block A was built at the front of the College. Thought by some to be for officers, they are believed to have stayed in the main building. Certainly it was used as a hospital ward, but no relevant records can be found despite diligent PRO searches. By 1942 when Grace Hardisty picks up the story the additional huts, including Block A, had been built; she never noticed them as she invariably drove at night up to the hospital main gates with young soldiers who had taken an overdose: 'neurotics' were the third largest group of illnesses treated at the military hospital.

Driving through the Blitz

Grace Hardisty[32]

It was in 1938 that I answered an advertisement for women drivers in Liverpool to be trained for the ARP Ambulance Service, though it was not until 1942 that I first drove an ambulance to Ormskirk Military Hospital at Edge Hill. In fact the first time I took a patient I thought they meant the hutments at the County hospital, but when I got there they were not expecting me, and said I should go to the 'New' Military Hospital. I never made that mistake again. I was based at Squire's Gate Training Camp, Blackpool where, with another driver, we manned a strict twenty-four hour on and twenty-four hour off ambulance rota. The Chevrolet ambulance, donated by the people of Ontario, was super, it was beautifully sprung, and had a big red cross on top, though it did only eight miles per gallon, and we had to go to Birkenhead to get petrol. Most of the transfers to the military hospital took place at night, and with the ambulance being left hand drive, I could see the hedges and pavements with even minute lights on dark nights, and could do the journey from Blackpool to the Ormskirk Military Hospital at Edge Hill in forty minutes. I would drive the ambulance up to the front doors at the main entrance on St Helens Road, and, if transporting POW, I left my co-driver inside the locked

---

[32] Telephone interview, September 2004, and copy of letter written to Dr Betty Underwood

ambulance, and called for an RAMC orderly to come. I only met male orderlies, though I didn't know any personally; I never socialised at the Military Hospital.

It was often 2.00am or 3.00am when I was called to transport young men who had tried to commit suicide, usually by overdosing on aspirin. Because of the light springing of the ambulance and the general condition of the roads, and the fact that it was necessary to get to hospital as soon as possible, usually by the time they arrived there was no need for stomach pumps, and the orderlies were despatched to clean and disinfect the ambulance.

After a posting to the Manchester area, from where ambulances went to Davyhulme hospital, I was posted as Senior NCO attached to the RAMC and working with the Medical Embarkation Unit on Princes Landing Stage in Liverpool, when I again came regularly to the Military Hospital, Ormskirk. Here I had eight drivers who manned three large awkward four-stretcher ambulances, two utility vehicles and two staff cars, used by the OCs, whose office was in the Liver Buildings. Also working from an office in the Baggage Room on Princes Parade was Captain Richmond, the Medical Officer for the docks. As a Liverpool GP, he would do his own surgery after a day on the docks. Converted Cunard liners and merchant ships, which acted as Troop ships, going on convoy to North America, and later to the Mediterranean, regularly brought in sick and wounded men, who were conveyed to Ormskirk or the Deva hospital, Chester. Thursdays were designated non driving days, to conserve fuel, when there were supposed to be no military vehicles on the road, but MO meetings were held at Ormskirk Military Hospital on Thursdays, and Captain Richmond always asked for a lift to Edge Hill.

Once Italy was invaded we brought German POW in need of urgent treatment to Ormskirk in the ambulances, and I carried a Red Cross card to avoid being taken as a POW in case of invasion. It was a fairly regular occurrence to go from Liverpool docks to Ormskirk Hospital. In early 1944 I recall the white painted Atlantis, a hospital ship, as against the camouflaged troop ships, coming alongside Princes Landing Stage. There were very many very badly injured and sick personnel aboard, and by this time Ormskirk Military Hospital was full and for this reason the orthopaedic wards at Alder Hey Hospital were utilised. In late 1944, after my marriage (I was allowed one week off) I was, to my disappointment, posted away from the Embarkation work to HQ, and so lost sight of the Port of Liverpool and the Ormskirk Military Hospital.

War, Dances and Romances

Several people in Ormskirk remember the **Saturday evening dances** at the Edge Hill Military Hospital. There were no adverts for these dances; the message was spread by word of mouth. Ron Tootle's two sisters, who lived in Wigan Road, Ormskirk, took a short cut across 'Muck Forshaw's' fields on their way to Saturday dances at Edge Hill. Both met their husbands there, one from London and one from Caerphilly. Lil Moulson remembers coming to collect her niece, who played the piano, from dances at Edge Hill. She left the Drill Hall in Ormskirk after the dance there had finished at 10.20, in time to let soldiers catch the last train back to Burscough, and walked in the dark, from the Ruff Lane entrance. Others remember the 'Skem Jazzer' trains to Ormskirk for the Saturday dances. It is believed that Bob Hope appeared on the stage of Hale Hall during the war, when he was stationed at Burtonwood air base near Warrington, the maintenance supply base for the US Air Force in Europe.

Dances and Romances

Beryl Reeve

Mrs Beryl Reeve[33] remembers meeting her husband, Robert, when he was a RAMC soldier in the '102' military hospital at Ormskirk (Edge Hill). Like many others he was stationed there for a relatively short time. He had left school in Beverley, East Yorkshire, at the age of fourteen and went to work as an errand boy, though he did evening classes in chemistry, to get on. He was in the choir at St Nicholas church, Beverley, and applied to be footman for the Bishop, but in fact got a job as footman for the old Duke of Devonshire at Chatsworth. The Duke had just had a stroke, and the nurse looking after him suggested that Robert might be interested in a nursing career himself. After the old Duke's death, the Dowager Duchess wanted to move house, and called for Robert, as she wanted him to accompany her to Hartington House. Instead he decided on nursing, and did mental nurse training in York and got his RMN. In the middle of his SRN training he was called up into the Army.

Beryl met her husband in 1943, when she accompanied her sister Margaret to a dance at Edge Hill. He was then a private, doing

---

[33] Telephone interview on 5 October 2004 with Beryl Reeve (née Draper) and 21 October with her sister, Mrs Margaret Goodwill, resident the Fountains, Ormskirk. Robert Reeve died ten years ago at seventy-nine, and Beryl is eighty-five and now living in Hull

guard duty on the Ruff Lane gate, and at first seemed reluctant to let them in until she said she was a pianist and part of the band. Margaret remembers an American soldier following them from the dance and asking to go home with them, as he had got too drunk to get back to Burtonwood, and to get rid of him the two sisters walked through the Fire Station on Derby Street, where the firemen were lying on the floor ready for a call. Margaret was a nurse at Alder Hey hospital in Liverpool, and during the war, when one of their patients who was transferred to the Military hospital left his pyjamas, she volunteered to deliver them, as she lived in Ormskirk, which she did, to the front door, but she was not allowed in.

Beryl had been a music teacher, but went into Hattersley's munitions factory as a store clerk during the war. After work she played the piano for dance bands and concerts, like the 'Battledress Follies' at the Regal Cinema in Ormskirk, now Tesco, and for two girls who danced, and also for Jean Gerard a local singer. Beryl 'got into' dance music in the war; she played with George Oliver's Band from Southport, and Bill Gregson's Band from Skelmersdale. Both bands went round the country villages, but did not play at the military hospital which had its own band, musicians and a choir, with people of all ranks, from privates to lieutenant colonels, and nurses.

Robert never discussed the patients in the military hospital, so Beryl knows little of his work there as a nursing orderly. They married on 13 May 1944, in the Parish Church, Ormskirk, and Padre Eddison, a Scottish man who was padre for the hospital, took part in the marriage ceremony. They had a week's honeymoon in the Lake District, at her aunt's hotel in the Lyth Valley, and then returned to Ormskirk. After their marriage they stayed in her parents' house in Cottage Lane, and he slept at the hospital only when on night duty. On the day Robert left the hospital in August or September 1944, after she saw him off at Ormskirk station, she was asked to go to the hospital to pick up some music, where she remembers going into an office and seeing Polish soldiers there. From Ormskirk, Robert went abroad to France, but was soon sent back to Peebles. He was then a Warrant Officer, and later went to Hull University, as a nursing tutor, as his war service helped his career. He later worked in Kingston General Hospital and Hull Royal Infirmary, working there until he was sixty-five.

As a gateway for transatlantic supplies, Liverpool was heavily bombed in the war. Lil Moulson remembers how when the docks were badly hit, the Tate and Lyle sugar factory exploded, blocking the railway, and an alternative bus service was provided from Kirkdale into the centre of Liverpool. From 16 September 1939, when the war began, Liverpool Corporation moved 85,000 children, families and teachers to rural Lancashire, Wales, and Shropshire, but forty per cent had returned by early 1940. A second programme of evacuation occurred after heavy bombing later in 1940: the Durning Road disaster was the worst single incident as recounted earlier. Over 40,000 people died in the Liverpool blitz, 10,000 homes were destroyed and 70,000 people were made homeless during air raids that peaked in May 1941[34]. Edge Hill Military Hospital sent staff to the County Hospital when the number of civilian casualties almost overcame their capabilities, while in return the County provided facilities for mental health care and for post-mortem examinations for the Military Hospital, and accommodated military staff.

The chief diseases at the Military Hospital, 1944	
Alimentary system (dyspeptics)	459
Respiratory system (bronchitics)	434
Nervous system	290
Rheumatic diseases	119

Further information comes from RAF research by Mark Gaskell, who has provided a copy of a war office record for the final six months, from 1 January to 21 June 1944[35], of the Medical Division Military Hospital, Ormskirk. This half yearly report noted that hospital administration had run efficiently in spite of the 'mass of forms and documents which have to be completed for each patient'. The report lists and categorises admissions to and transfers from the hospital, and mentions that, because from January 1944 the hospital 'has been a centre for treating cases of amoebic dysentery invalided from overseas', probably Egypt, and waiting for TB cases to be transferred to sanatoria, on several occasions they had to refuse admission to any but urgent cases. Appointments for a maximum number of twenty outpatients per day had also been set. Overseas inpatients were from convoys from 'all theatres' (of war; Mediterranean, European, Atlantic and Asian) and there were many recurrent cases of skin disease, neurotic and psychotic illness and pulmonary TB. Home force in-

---

[34] The Spirit of the Blitz, Mersey Maritime Museum, September 2004
[35] File WO222/844 in Public Record Office, information from Mark Gaskell, 15.09.04

patients were divided into four groups; dyspeptics, rheumatics, bronchitics and neurotics and psychotics. Total hospital admissions during the six-month period were listed in the appendix as 1652 patients, though this number had been annotated to 1854, including both male and female officers.

There is some confusion in people's recollections of when the military hospital at Edge Hill closed, but a hand written note on a file in the Public Records Office states that the hospital closed down on 21 June 1944. Lil Moulson remembers that two (RAMC) units left for VE day (Victory in Europe Day, May 1945) though no one was told where they were going, and then the Polish (Army) came in.

Fig 3.14 Polish graves in St Anne's churchyard, Ormskirk.

Source: Mark Gaskell

No 4 Polish Military Hospital is listed as being in Ormskirk, and probably moved into Edge Hill in June 1944; Beryl Reeve saw Polish soldiers there in August or September 1944. Mona Duggan's second edition of *Old Photographs of Ormskirk* includes a photograph of Joyce Welsh, whose family lived in the Lodge during the war, outside Edge Hill with a smiling Polish soldier. The Polish soldiers were said to be very fond of children as they were missing their own families. In November 1944 Ormskirk UDC mentioned 'considerable assistance has been asked for by the **Polish hospital**'. The big problem was finding accommodation for Polish Officers, relatives and friends, while the Polish soldiers have nowhere to spend their leave. They wondered whether the Patients at the Hospital should have some instruction in English, and three teachers were organised for Polish soldiers. A visit was made to the Polish hospital at Christmas, while on 2 March 1945; Polish soldiers had distributed **chocolates** and sweets to the two Ormskirk schools, and in return 5600 **cigarettes**[36] were bought for them for £26 9s 7d. This may tie in with the twenty-five Polish graves in St Anne's churchyard, Ormskirk (Fig 3.14). The Polish hospital was only in Ormskirk for a short time, as records show that they moved to Iscoyd Park, near Whitchurch later in 1945. The dances did not apparently continue with the Polish Army. Returning Edge Hill students remember seeing Polish soldiers cultivating the fields in front of the College. Until early 2005, a wartime morgue still existed behind the Rose Theatre, whose stone slabs were complete with drains for body fluids[37], though it is not clear if this morgue was ever used. Derek Sumner, Head Gardener, reported in 2004 having found bits of Army uniform around the morgue in the 1960s. Little more is known of this 1939 to 1946 period, when the staff and students of Edge Hill were over the Pennines in Bingley in West Yorkshire. It is to this story that the next section refers.

---

[36] It is interesting to note that smoking was considered harmless at that time
[37] The morgue was demolished in 2005 to make way for the new Performing Arts building

4: The Bingley Years: 1939-1945

Moving house is said to be one of the most traumatic episodes in modern life: moving an entire College with its staff, students and resources must have been so, yet all records suggest that this move, and eventually back again to Ormskirk was carried out with good humour and efficiency. This is what happened to Edge Hill in September 1939; the Army requisitioned the Ormskirk site and the College moved to share the accommodation of Bingley College in

Fig 4.1 Bingley College, West Yorkshire, shared with Edge Hill in WWII (shown on the map as West Riding Training College)

Source: NGAS Map Collection

West Yorkshire (Fig 4.1). It may have exhausted the Principal, Miss Smith, who only in 1933 had overseen the College move from the old buildings in Liverpool; failing sight and poor health forced her to retire after only two years in Bingley. Traumatic events make for powerful memories, and these form the basis for this chapter. 'Bingleyite' Edge Hill students, Constance Bancroft, Alice Dale and Elma Eastwood present their clear and vivid memories of Bingley and their feelings towards the place engendered by the difficulties of studying geography in wartime conditions. Emily Brontë's 'regional novel' *Wuthering Heights*, written at nearby Haworth, is an illustrative backdrop for conditions in the Bingley area during the war years, although students at this time did not use this kind of literary association in their geographical studies[1].

It was in May 1939 that Miss Smith, Principal of Edge Hill Training College, had been informed that in the event of war the college would be requisitioned for the Army, and preliminary arrangements had been made to share the College at Bingley. She had no confirmation before the summer break, so it was at the end of August 1939 that she had to organise a rapid evacuation. Mobilization at the beginning of September brought soldiers immediately into the Ormskirk building. It was inconvenient; had the students known they could have moved directly to Bingley after the summer break; belongings stored in halls over the summer had to collected in September. As it was special transport had to be ordered, and all their belongings repacked for the new move to Bingley in West Yorkshire on 4 September. Most of the College furniture was sent to Widnes to be stored.

Bingley lies on the River Aire, close to Bradford, by which it tends to be overlooked. J B Priestley writing his *English Journey* in 1937 did not visit or mention Bingley, though he travelled by car from Nottingham via Chesterfield and Sheffield to Barnsley and Huddersfield to Bradford, which although he had lived in Bradford up to 1914 he described as 'a city entirely without charm'. He saw the area characterised by moors, stone walls, and woollen mills: Bingley was part of the Yorkshire woollen industry in the Industrial Revolution, and had many woollen mills. Priestley saw *Wuthering Heights*, written by Emily Bronte in 1847, 'just around the corner', and the 'hills and moors and dales are there for you'. By 1937 Priestley noted that 'clogs and shawls are disappearing' from Bradford. The hilly site of **Bingley College** was on Lady Lane (Fig 4.2). Miss C. Fletcher, Principal of Bingley College between 1936 and 1946, remembered the time when Edge Hill came. The two Colleges doubled up for the war, working together yet keeping separate

---

[1] Literary associations, such as Wordsworth or Beatrix Potter in the Lake District, form an important part of modern human geography

identities. It was very difficult feeding over 300 students in each of four small dining rooms, with an inadequate kitchen and rationed food. Students cleaned their rooms and sometimes had to prepare food. Air Raid practices were held, but only a few firebombs fell in the grounds over one vacation. The blackout was a headache in a College on a hill with a thousand windows[2], and delayed the 'Welcome Party'. Bingley College also had to fit in an extra twenty students for a few weeks when Hull Training College was bombed. Miss Smith wrote 'such circumstances call for all the courage and endurance of every one of us...'[3] though lectures, concerts and productions of the Musical Society did continue. There was also a talk on 'The use of Wireless for Schools', when the speaker said the BBC should firstly give information (about the war no doubt) secondly education and thirdly entertainment.

Wartime blackout
It was feared that bright lights would prove an easy target for enemy bombers, so from 1 September 1939 major cities were ordered to blackout any lights that could be seen from the sky. Air Raid Precaution (ARP) Wardens supervised blackout. People were required to cover windows with anything that could stop light escaping, paper, curtains, blinds, even paint. Car headlights were taped up except for a small slit.

On arrival Edge Hill students had to share bedrooms in the **halls,** doubled-up in single rooms, as everyone was, part of our 'war effort'.[4] Priestley Hall, named after JB Priestley, housed Stanley and Clough, while Acland Hall[5] housed John Dalton and Lady Margaret, though some Edge Hill students were in all five of Bingley's halls, Acland, Priestley, Aschan, Alguin and Hild (See Fig 4.2 lower). Classes were not shared; each College timetable had to be fitted into classes from 9.30-12.00, 1.00-4.00, with 5.00-6.00 instead of evening lectures, presumably because blackout requirements and low levels of lighting made evening study difficult, though students were expected to carry out private study at night. Morning assemblies alternated between the two Colleges, though each had a piano. Fiona Montgomery (1997) notes that conditions in Bingley during the war were not much better than those of early Durning Road, Liverpool days.

---

[2] Edge Hill Guild Newsletter 2001 p20, extracts from Bingley College Jubilee Magazine 1961
[3] Edge Hill College Newsletter 1941, No 50
[4] Letter from Constance Bancroft, enclosing the Bingley College postcards, inscribed Margaret Singleton 1943-45 though borrowed from a friend in Canada
[5] Acland Hall may be named after James Acland (1799-1876) a political radical and agitator from Hull

Fig 4.2 Postcard views of Bingley College. Upper: The College block is on the left, with the Halls of Residence uphill across the grounds. Lower: The Halls of Residence, from left to right, Hild, Alguin, Ascham, Priestley and Acland.

Source: Constance Bancroft

Not all students were fully committed to their studies, as some girls apparently chose college to avoid being called up; girls were liable at eighteen. School Practice was a great problem at first as Bradford schools were closed, so practice was found at other Aire valley schools, including in Skipton. The final practice was held in most severe weather, when students had to walk long distances in ice and snow. Most students did some war service in their vacation either on the land or in hospitals or with evacuated children. Other students had helped a boys' school bring a piece of land into cultivation in the hospital grounds.

Wartime Geography

During the Second World War many geographers were recruited as meteorologists, because of their skills in working with maps. Their main work in weather forecasting was the preparation of synoptic charts. They prepared forecasts for bomber operations and advised on the choice of sites for the Normandy landings (O'Hare in Taylor 1951). The call up of many geographers left some schools short of teachers, while from 1940 paper was rationed. The 1940 GA conference was held in Blackpool instead of London, and there was no population Census held in 1941.

However, during Edge Hill's time at Bingley, normal Edge Hill teacher training occurred, students took one subject at Advanced level, and three at Ordinary Level. Miss CM Giles[6] (1939-1941) remembers learning **geography** at Bingley. Miss Butterworth, who had taught geography at Ormskirk, taught her. The geography appears to be as set out by the Board of Education, covering regional geographies of Britain and the world; 'we studied maps of the British Isles and learnt about the hills, rivers, and coastlines. Then we had maps of the world and learnt about other countries, the climate, the people and industry'. Miss Butterworth, like other geographers of the time, based much of her teaching on maps. Students were encouraged to buy their own atlas, and her field visits involved following the route on an OS map. Teaching used Bingley College's geography room and may have shared their resources, which were probably in short supply as Edge Hill's library didn't arrive at Bingley until 1943. Other students remember a demonstration lesson in a Bingley school by the Geography tutor (unnamed) who 'described the Indian countryside, mountains, plains, animals and life of the people. Students drew the animals and the best were put on a frieze',[7] reflecting the continuing dominance of imperialism and the need to know the geography of the colonies. As in Ormskirk geography was taught in the Education block in the main College building (see Fig 4.2 upper). 'From the Hall

---

[6] Questionnaire response August 2004
[7] Edge Hill Guild Newsletter 1989, p.25

of Residence we walked across a field to the education block. There was a large hall, and several smaller rooms. We used one of these' (C M Giles 1939-1941).

Fig 4.3 Geography Expedition to Ingleborough 1940

Source: Sylvia Woodhead

Perhaps because of a shortage of teaching materials, or perhaps because of the interesting locations surrounding Bingley, Edge Hill students seemed to do more geography **fieldwork** while based at Bingley, despite wartime restrictions on travelling. 'Is your journey really necessary?' was the message on many wartime posters. 'We walked on the hills and followed the River Wharfe. We went on a tour of the Shipley canal and saw a lot of the countryside'. However 'It was war time and we spent a lot of time in air raid shelters during bombing raids' (C M Giles 1939-1941). Jennie Parson (née Styer 1941-1943) remembers a fieldtrip to the limestone scenery of Ingleborough Cave and Gaping Gill,[8] (Fig 4.3). The 1940 Edge Hill College Magazine notes in great detail the Biology and Geography expedition to Malham and Gordale, when Miss Newton, Miss Jackson and Miss Butterworth led forty-eight students.

Expedition to Malham

> We left College at 9.00am… by two motor coaches which conveyed us to the picturesque village of Malham, situated

---

[8] Edge Hill Guild Newsletter 1989, p.35-36

in the Aire Gap, passing en route the towns of Keighley and Skipton. There was much to notice on the way, the broad Aire Gap, the effects of former glacial action, the change in scenery from the sandstone to the limestone moors. On arriving at Malham at 10.15am we proceeded on foot by Gordale Beck to Janet's Foss where a succession of waterfalls, pools and rapids was noticed. Owing to the recent drought the falls were dry and we could see the interesting formation of tufa on the sides of the waterfall and even penetrate behind it. Evidences of undercutting and solution were noticed in the formation of caves. From here we walked… to Gordale Scar, the sides of which closed in overhead, showing the former presence of a cave whose roof had collapsed. This was a most striking spectacle and again we could see masses of tufa in front of the waterfall, which was however dry. The springs in front of the gorge were dry and the beck was very low. Evidences of the fault line could be seen in the presence of both limey and sandy soils… After lunch we explored the pavements of Malham Cove which showed the formation of clints between which soil had collected… We climbed down the steep hillside into the cove where one of the headstreams of the river Aire comes almost imperceptibly from the base of the limestone cliff… from the cove we divided into smaller parties, some to explore and some to have tea, to meet again for the homeward drive at 5.30pm. The weather has been kind and the country was very beautiful. (Report by E. Green and N. Chadwick, two students).

Despite the fear of bombing raids, and wartime restrictions, Edge Hill students did enjoy their Bingley days, which bring back many strong memories of the place. Ruby Demain (1939-1941) writes about the very severe winters at Bingley. 'Snow lasted for such a long time - going home on the bus for Half Term between huge banks of snow dug out by gangs of workmen. Keeping the blackout blinds down in a gale was a work of art. Lights were very dim anyway'.[9] Amy Morrissey (née Williams) now in Arizona, had 'enjoyed her year in Bingley more than her time at Ormskirk. The surrounding countryside was much more interesting and beautiful, especially the moors'. Alice Humphries (née Holt) was also happier at Bingley, again because of the surroundings. Many students remembered the walks over the moors to Haworth.[10]

---

[9] Edge Hill Guild Newsletter 2002, p.29
[10] DM Fox (nee Elliot 1938-40)

In 1941, while at Bingley, Edge Hill College had its Jubilee Year. Miss Smith retired due to ill health, at the end of the summer term, and Miss Butterworth took over as Principal, possibly leaving a gap as geography teacher; the Geography and Biology expedition to the nearby Washburn Valley, Leathley and Riffa Wood merited only one sentence in the College Magazine, which also noted that all students had taken a first aid course and there had been fewer social gatherings. The very thin 1942 Edge Hill College Magazine was written on the third year away from Ormskirk by E M Butterworth, who noted that 'students had travelled to Haworth and Bradford' and that 'Miss Newton is taking over classes in Gardening and care of the grounds', while 'most members of the College are engaged in some form of service, from fruit picking to munitions work or from hospital ward to war nurseries'. Betty Stewart (1942-1944) recalled many memories of Bingley,[11] while Constance Bancroft who was at Bingley from 1943-1945, regrets that she can remember nothing about the geography she learned there except that the transmission of malaria is by *Anopheles* mosquitoes. She contacted Kathleen Haworth (née Minto) who also took geography between 1943 and 1945 and remembers only lianas!

Fond Memories of Bingley

Constance Bancroft 1943-1945

Your letter made me realise I cannot remember who was in the geography group. I would say there were fifteen to twenty of us. I can remember much more about school geography. We had four years of that in the 'express stream', with two excellent teachers, one of them the Head until he went into the Army. I got a distinction in School Certificate and would have done it for Higher, but it was not on offer in our rather small, rather second-rate grammar school with several male teachers 'called up'. The Bingley College was far better placed, in my view, than Edge Hill at Ormskirk. The river Aire runs through the town, with (at that time) woods and open country on either side, and we could cross the Leeds-Liverpool canal on the way to church. There were several walks on the moors and for a day out we could walk over to Ilkley and get two buses back.[12]

Bingley College was situated high above the town, out on the road to Bradford, up a steep hill called 'the Struggle'. From the Halls

---

[11] Edge Hill Guild Newsletter 2004
[12] Questionnaire response August 2004

there was a view over the valley; I remember seeing fields with snow, though you couldn't see Bingley. We had geography in the same room after the Bingley students and we saw Bingley's examination questions written on the blackboard (because of paper shortage) but didn't think much of them - we had done similar at Advanced Level.[13] The sharing with Bingley College was on a strict timetable and went like clockwork. Edge Hill students didn't mix much with Bingley students, until the war ended when we had a joint concert.

I have fond memories of Miss Butterworth, the Principal, who must have been a raving beauty when younger. We were all girls and liked to dress up for dances. Miss Butterworth had a long blue dress to match her beautiful blue eyes. She was not very tall but striking, especially when she wore her white hair 'on top', instead of the normal bun at the back of the head.

Fig 4.4 Bingley College Halls of Residence (now a Nursing Home) in 2005
Source: Kate Chapman and Daniel Tetlaw

We didn't see much of Miss Butterworth, except when we were called to her room at the end of the year for a progress talk, when she was very perceptive, and knew who shared rooms with whom. I can

[13] Telephone interview, 25 November 2004. Edge Hill was an 'A Class College', implying that Bingley College had a lower grading

remember standing at the bus stop for my very first (Infant) teaching practice in Bradford, with a rabbit in my arms to interest the children, and Miss Fletcher, Principal of Bingley College, happening to pass, said in her erudite Oxford accent 'I see you are taking your livestock with you'. Miss Butterworth was much more down to earth, speaking with an educated Liverpool accent.[14]

In 1943, the fourth year of evacuation, Miss Butterworth commented that 'Miss **Dora Smith** had already proved a keen geographer, taking expeditions to York and Malham'. Alice Dale (1944-1946) has clear memories of one of these Malham excursions.

Excursion to Malham

Alice Dale, 1944-1946

Two day excursion to Malham led by Miss Smith - Geography and Miss Crennell – Biology. By train to Bell Busk, standing the whole way, looking out for the all the things Miss Smith had mentioned. Panic at Bell Busk Station as six of the party were missing. As the train moved off they were seen shouting and waving. They were trapped in the last coach. The Station Master leapt on the train and rescued them.

We walked to Kirkby Malham, Miss Smith lecturing all the way in a very school-marmish manner. At Kirkby Malham we were met by the Rector's wife and had some sandwiches and tea on the lawn. On to Malham Youth Hostel, Miss Crennell getting in a word wherever she could about the flowers in the hedgerows. We left our luggage at the Youth Hostel and went up the road past the Tarn, the way that Dorothy Potts (now Firth) and I had climbed the previous Sunday when we had cycled from College (in Bingley) and spent a night at the hostel. We took a different path down to the tarn and found two new specimens in the bog, butterwort and bird's eye primrose, back across a moorland path, over some clints, where there were new specimens in the crevices and down a huge dry valley to the cove.

After our evening meal Margaret Powell, now Carmyllie, Betty Foster and I went up Gordale Scar and down a dry valley. Back to the hostel at 10.30, sixteen of us in one room. Breakfast at 7.30 and a walk to the waterfall Janet's Foss and on to the scar. After that we split up. Our group had to go to a farm and ask all the questions I had done

---

[14] Telephone interview, December 2004

Fig 4.5 The Main Entrance of the former Bingley College in 2005

Source: Kate Chapman and Daniel Tetlaw

when doing my Thesis 'Farming in the Lune Valley between Halton and Formby'.

After lunch more group studies, a walk through the water meadows to Airton studying the flowers and afterwards making a plan of the village. We had a cup of tea at the inn in Airton and set off to walk back to Bell Busk. A lorry stopped and gave some of us a lift. Suddenly it stopped again and two of the men jumped out, one shot a rabbit and the other ran for it, they leapt back in and off we went again.

All this from a letter to my parents. My mother kept all my letters and I have enjoyed re-reading them, especially the V. Day one when Dorothy and I tried to cycle to York Minster for the service but turned back at Tadcaster, soaked to the skin.

Examination results April 1945	
Geography	
Ethel Eccleston	60% (top marks)
Alice Dale	58%
Margaret Powell	50%
Several others around 50%, one failed.	

I am surprised at the low marks. Apparently 30% was the pass mark and 60% excellent. I had kept my College notes, my thesis and my wild flower collection but lost them in a flood in my home.

Excursions such as Alice Dale recounts were annual features for Edge Hill students while at Bingley, even for those for whom geography was not their main subject. Dorothy Firth (née Potts 1944-1946) was one of these; however her trips to Malham and to the locks at Bingley stood her in good stead when organising her own school trips. She taught for some years in Skelmersdale where many of the children had not seen hills. Dorothy still has a friend she met at the sanatorium while at Bingley. They started teaching at the same school and have been friends ever since, and keep in touch even though she has emigrated to Canada.[15]

The 1944 and 1945 Edge Hill College Magazines are very brief, Miss Butterworth commenting 'I shall miss the hills'. The 1945

---

[15] Questionnaire response, August 2004 and telephone conversation, 22 December 2004

session began in Bingley. Second Year students did School Practice in Leeds, Pudsey and Keighley, and First Year Students had a fortnight in Bradford and Shipley before moving began. Students heard a lecture on the development of Dr Barnardo's homes, and also attended a lecture of the work of the Tennessee Valley Authority, and Miss Billington spoke on the project method of teaching as practised in her school in Leeds. There had been expeditions to Malham and another to Fountains Abbey.[16] Elma L. Eastwood (née Bownass 1945-1947) started her College days at Bingley Training College.

A Study of the Heysham Peninsula

Elma Eastwood, 1945-1947

I took geography at Advanced level, and can remember studying geology and the geography of the USSR, with Miss Dora Smith. We had to do a geographical thesis of the area where we lived as part of the course and this was done in our summer vacation-this was our only geography fieldwork, as it was the end of the war. As I lived at Heysham I did a study of the Heysham Peninsula, which I still have today. I spent my vacation visiting the Preston Records Office to look at old maps of the area showing settlements, population statistics, land utilisation, railway development. We had to produce a rough copy of all our findings when we returned to Ormskirk in September. I also visited farms in the area, a bobbin mill, ICI, the harbour and Holiday camps – a very busy vacation. After our work was vetted by Miss Smith we then had to write it all out neatly and trace and draw all the necessary maps-painting and tinting them- no computers or photocopiers in those days.

The 1946 Edge Hill College Magazine was written after the return to Ormskirk, noting that Miss C. Fletcher, Principal of Bingley was to travel in West Africa on a Colonial Research Fellowship. H. Stevens, a student, wrote in the 'Exodus from Bingley' page that 'after half term in the autumn of 1945 'when we go back to Ormskirk' seemed much sharper in meaning. We packed and left in the last week of term. Bingley College organised three leaving parties, and the Colleges exchanged cheques as presents. The removal vans came and went quite irregularly'. They had shared the buildings for six years, during which time Edge Hill students had not known Ormskirk.

---

[16] Edge Hill College Newsletter 1946

After the war Bingley College fared less well than Edge Hill, despite similar isolation and a boarding school atmosphere,[17] and it suffered a series of mergers. In 1978 it combined with Ilkley to become Bingley and Ilkley Community College, and later with Shipley, and in 1982 it merged again with Bradford College; its higher education provision is now part of Bradford University, where some archives may remain. Returning Edge Hill students have discovered that the main Bingley College buildings (Fig 4.5) and halls of residence have been converted into flats and retirement homes (Fig 4.4) and feel a sense of loss that their 'Bingley years' memories have no college home.

---

[17] One student didn't like her time at Bingley College from 1947-49

Part II Modern Times

5: Post War Austerity in Ormskirk 1946-1960

After the war Edge Hill returned to its buildings in Ormskirk, and training of teachers was continued in the face of post war changes to society and education. It was the time of much welfare state legislation, but also post war shortages. This chapter follows the experiences of those returning to the Ormskirk campus after a gap of six years: Dorothy Firth describes how delighted students were to return to Ormskirk. Gladys Cooke remembers the strict life of being an Edge Hill student, while Lois Ford's copy of College regulations and exam papers reveal much about 1940s geography teaching at Edge Hill, and Olive McComb reflects on living in Block A. Army buildings, including what is now the geography block, had been constructed, and needed to be converted into educational uses. All the furniture of residential living, the library books and teaching resources had to be returned and replaced. Students found themselves roped in for domestic duties. There was the possibly unwelcome discovery that no one was prepared to be a resident maid any more[1]. Gradually normality was resumed, and the grounds restored to something like their pre-war condition. The College continued to grow and a building programme was begun. Similarly geography began to develop from its previous imperialistic stance. This chapter provides a link between wartime and modern degree provision.

Miss Butterworth, the Principal, had visited the Ormskirk campus from Bingley late in 1945, when she saw that the floors were black, there was no kitchen machinery and the whole building needed re-decorating. Edge Hill staff and students returned to Ormskirk 'our own restored buildings' to quote Miss Butterworth in January 1946, accepting fifty-one extra students, making 280 in September 1946. Return of the furniture from Bingley and the store in Widnes was not easy, and everything had to be washed. Miss Butterworth notes that 'the buildings put up in the grounds (including Block A) have been acquired by Lancashire and are being adapted to make student rooms (Block A) classrooms and a Hall to take resident domestic helpers, as soon as we find any' (Fig 5.1). This latter proved a vain hope; the days

---

[1] Miss Bain, later to be Principal, commented in the Edge Hill College Magazine No 64 1955, that there was no room for maids, as student numbers rose to 280, compared to 180 before the war, although there was the extra accommodation provided by the blocks and huts left by the Army

of personal servants were over. She noted that Miss Newton had returned from Kent to supervise the rehabilitation of the grounds, but does not mention any geography staff, except that Miss Dora Smith (later Mrs Whitworth) had come to a reunion. On their return First and Second Year students took turns at domestic work. Before Miss Butterworth retired, she had Block A painted white, since when it was called the 'White Hut'.

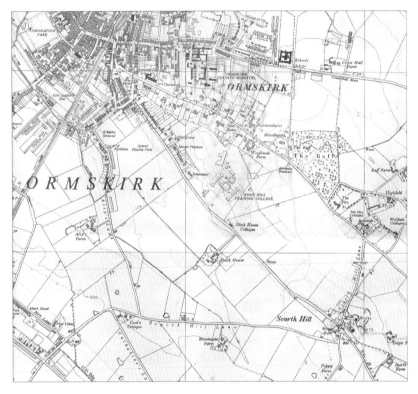

Fig. 5.1 Edge Hill's post war campus. The Geography Block can be seen adjacent to St Helens Road: Extract from OS 6" sheet SD 40 NW 1955

Source: NGAS Map Collection

'Anyone approaching College from either entrance will now see, instead of the brick extensions, which were inharmonious with the main buildings, charming dark-roofed, cream coloured buildings showing most attractively against the greenery of trees, grass and shrubs. Inside them, good-sized study bedrooms have been fitted up and the furniture successfully treated to bring harmony and brightness into the original motley of supply[2]'.

As a geographer Miss Butterworth was very interested in the College grounds, and the regular **'In The Grounds'** reports resumed in Edge Hill newsletters. In March 1946 work began to the shrubbery and ornamental gardens, though it took several years to get the grounds back into order after the war.

Fig. 5.2 An early photo of Swift's Cafe in Ormskirk

Source: Mona Duggan

> **1947**: 'There is much still to be done, but wonders have already been achieved in restoring the grounds. We have had fresh produce all the year.... The students' plots were many and particularly good. The Rock Garden has been re-sown and the 'islands' have emerged from the over-growth of years... We hope soon to rehabilitate the orchard that Miss Smith so kindly gave, for it suffered very badly during the war years. The Students' Avenue is doing well and at this moment its small rowans are heavy with reddening berries... It is lovely to have such space and beauty.'

Mabel Southworth[3] (née Clarke 1946-1948) remembers that to the left of the entrance the hockey field had been ploughed for crops

---

[2] Edge Hill College Magazine 1947
[3] Edge Hill Newsletter 1998

and Italian prisoners of war were still working there in 1946, although the College had been most recently used as a Polish military hospital, its legacy said to be bed bugs in study bedrooms. Many corridor floors were badly damaged and there were few lights.

None of the students knew their way about. After dinner Mabel Southworth visited the 'superb Swift's café in Ormskirk High Street which served beans, chips and orange or chocolate cake' (Fig 5.2). Following this, Edge Hill students went to the cinema, 'Pavilion or Regal for 9d each'. They were appreciative of the lime trees and peace in the gardens. Games could now continue: hockey and netball in winter, with rounders and tennis in summer. In the summer term of 1946 visiting lecturers included a Ministry of Information films programme, while in July 1946 the Advanced Geography and Biology students visited Loggerheads in North Wales. Some of the first cohorts of students to return to Ormskirk after the war remember Edge Hill experiences. Dorothy Firth (1944-1946) and Elma L Eastwood (née Bownass 1945-1947) both started their Edge Hill course at Bingley, then returned to Ormskirk, where 'much was in chaos and in turmoil' on their return.

Excited to return to Ormskirk

Dorothy Firth

We were excited at the thought of returning to Ormskirk. We wondered what the place would be like. Our common room had been used for operations and had to be fumigated. We had to walk on planks as many of the floorboards had been dug up. The campus was heavily used in the war. Attitudes were very different then; nobody complained. All of us mucked in to do what we could. People who haven't experienced it have no idea what the wartime spirit was. People were very helpful and supportive of each other. There was no fear about going out, and although the food wasn't good we had enough to eat. We were thrilled to have a room to ourselves after sharing at Bingley, and thrilled to have a swimming pool[4].

E M Butterworth: Student to Principal

In 1948 Miss Butterworth retired and Miss Margaret Bain (Fig 5.3) became the new Principal of Edge Hill, and continued both the Principal's report and **'In the Grounds'** in the College Magazine. It is

---

[4] Telephone conversation 22 December 2004

Miss Butterworth: a life

Born, 11 March	*1882*
Educated at Pupil Teachers College, Leeds	*1898-1902*
Edge Hill University student, Durning Road Liverpool.	*1902-1905*
Liverpool BA & Board of Education Certificate (of teaching) Class 1	*1905*
Teacher at Jarrow-on-Tyne Pupil Teachers Centre	*1905-1907*
Teacher at Lincoln Training College	*1907-1916*
Student at St Hugh's College, Oxford	*1916-1917*
Oxford Diploma in Geography with distinction	*1917*
Teacher at St Stephens High School, Clewer	*1917-1920 & 1920-1922*
Student at Universities of Paris and Grenoble	*1920-1921*
Teacher at St George's Ascot, Sept - Dec	*1923*
MA in English, Liverpool University	*1923*
'Came to live with us' at Durning Road	*1923*
Teacher at Polam Hall, Darlington, January-April	*1924*
Teacher at Ware Grammar School, January-July	*1925*
Teacher responsible for Geography, also teaching Education, at Edge Hill	*1925*
Published Elementary school geography textbooks, 'From Pole to Pole, The Overseas Empire, The Old Country' in the Geography Why and Where Series by Blackie	*1925*
Geography Expedition to Windermere, by coach	*1929*
Visits to Cammell Laird's shipyard & Llangollen	*1930*
Expedition to Windermere & Kirkstone Pass	*1931*
Expedition to Loughrigg, Grasmere & Rydal Water	*1932*
Expedition to Ullswater, by train, bus & steamer	*1933*
Became Editor of Edge Hill College Magazine	*1933*
Moved with Edge Hill College to Ormskirk	*1933*
Led Geography Expedition to Langdale	*1934 & 1935*
Led Geography Expedition to Keswick	*1936*
Geography Expedition to Wansfell and Troutbeck	*1937*
Expedition to SW Lancs coast, Hightown	*1938*
Led Geography Expedition to Ambleside	*1939*
Moved with Edge Hill College to Bingley	*1939*
Geography Expedition to Malham & Gordale Scar	*1940 & 1941*
Became Principal, while at Bingley	*1941*
Returned to Ormskirk	*1946*
Retired	*1947*

interesting that recent tree planting has endeavoured to match the early planting described below. The sweet chestnut and flowering cherry trees in front of the main building are now imposing mature trees; their size has surprised returning students who remember them much smaller.

> **1948**: The grounds have now, as a result of unstinted effort and hard work, resumed much of their beauty and have been a source of usefulness and delight... The summer house has been moved to face south... The Rose Garden is as lovely as ever, although much replacement of roses is still needed... Dr Bain has made a gift of twenty birches in the 'Smith Orchard' and round the new buildings.

> **1949**: This year we have been very tree conscious. We have bought trees, been given trees... and had the keenest pleasure in admiring those which were planted long ago and are now nearing maturity. It is interesting to remember that there were only three trees on this site of thirty five acres when we came to Ormskirk. They were the great sycamore and two elms by Mr Alcock's farm. Yet trees grow here to perfection when they are sheltered from winds. The Students' Avenue of hawthorn and mountain ash trees is doing very well. The first year they were planted (1935) seventeen were blown down in one night, but with age and more efficient staking, they have withstood the gales, and most of them have flowered well. The lime trees in Ruff Lane entrance are now very beautiful... Those which were planted at the same time along the main drive never succeeded, and they were replaced last year. The trees in the Rock Garden are doing very well. The beeches and birches now form a wonderful back-ground. The specimen trees on the lawn are good - the weeping elm being a special favourite. The Sweet Chestnuts are growing to perfection... The two acacias, two cedars of Lebanon, the almonds and flowering cherries, and many pines are doing well. Perhaps the greatest improvement of all this year has been the re-sowing of the great lawns. The Gardens have received many gifts this year. One Old Student of Bingley days gave six flowering cherry trees.

Geography in the 1940s

In the late 1940s some geographers still exhibited a masculine gaze; A F Martin, writing in 1951 on 'the necessity for determinism' in geography, begins with the sentence 'geographers are on the whole fairly practical *men*' (my italics). **Determinism**, the idea that people are limited in their endeavours by the natural environment, still survived, shown notably by the Australian geographer Griffith Taylor, who in 1951 published a volume on twentieth century geography, with contributions by twenty-six geographers. Griffith Taylor considered geography to be concerned with the older ideas of environmental determinism, exploration and regionalism, while Harrison Church argued the case for colonial geography; there were still such courses in some French Universities (Harrison Church in Taylor 1951). Physical geographers like Wooldridge argued that we should not be ignorant of the (British) Empire; he identified the need for geographical studies of the colonies (that is the tropical zones) to help in development. 'Maps are required of the distribution, intensity of production and marketing routes of most colonial crops'. Griffith Taylor took a determinist view of the environmental controls affecting the distribution of heavy industry and assumed that weapons production depended on iron and steel supplies. He produced 'habitability maps', for example of Australia, with 'isoiketes' as lines of equal population, though he did not include plastics and atomic power in his calculations as he considered their uses too unknown. Ellsworth Huntingdon, author of *Civilisation and Climate* in 1915, considered in 1947 that the ability, created by air travel, for world leaders to meet would prevent any future wars, and said 'in a few years we shall have millions of people who know some foreign country personally' (in Taylor 1951:53). Huntingdon did not think aircraft would be able to fly across mountain areas, but believed that aviation would increase world contrasts. He assumed great public ownership of helicopters, which he saw as important in commuting, and in causing the spread of cities.

P K O'Hare noted that advances from the recent war now dominated geography. New knowledge of air masses gained from high flying aircraft had led to the use of new terms in meteorology, like fronts and Rossby waves, and that now radar could be used to detect rain storms, while aircraft had been used in rainmaking experiments. Exploration was still important, particularly of the North Pole, where even the position of the Magnetic North Pole was uncertain, and the Antarctic where much of the interior was still unknown. Isaiah Bowman referred to pioneer settlement of 'frontier zones', in Canada, Australia, interior Brazil and Northern Rhodesia, the latter where 'the railway is the base-line of settlement... Its construction is a matter of colonial policy and a link in the Cape to Cairo chain of transport and a factor in imperial economic interest and defence' (in Taylor 1951:256).

The dramatic changes in society and geographical knowledge following World War II seem at first not to be evident in immediate post-war life at Edge Hill, as recounted by two students, who relate their life there in the 1940s.

'They were good days'

Gladys Cooke 1946-1948

I was only seventeen when I started at Edge Hill in September 1946. I remember Miss Butterworth organised a competition to see who could dig up the longest dandelion root, a clever way of getting students to weed the grounds. To me the grounds and College seemed to be in good shape, despite Miss Butterworth's complaints. We did gardening and had a plot at College where we grew vegetables. I enjoyed this as a link with home. I didn't take the geography course, but think they were taught the 1933 syllabus. All the tutors seemed very old ladies. They were not well liked, they were very strict, forbidding and old fashioned, not understanding of people. They seemed not to have families, and to be sad people. Miss Dora Smith, a very small person, was very strict and severe, and Miss Butterworth, the Principal until the end of 1946 was particularly horrid and staid. Miss Bain, the next Principal, was more gentle and understanding, but still not very approachable. The most cheerful was Miss Cook who taught Maths, though Miss Stevens was nice and caring.

We had to be in by 10.00pm, if not there was severe trouble. We were not allowed to speak to young men or to have any men, even our father, to our rooms, though female relatives were OK. We had to book the downstairs Coaching Room to meet a 'gentleman'. I was still a schoolgirl really, so I didn't mind the restrictions. I remember lectures in the tiered lecture theatre in the main Education block, and modern dance in the Hale Hall. We had to get the PE uniform of green blouses and shorts from Leeds, and wore sleeveless blue dresses for Dance lessons. We also wore ties and College green blazers, with yellow and purple braid. There was an Art room at the end of the main block where Arts and Crafts was taught. No one liked approaching it at night, as rumour had it as the mortuary for Polish soldiers. All first years shared a room, with ex WRAF/ WREN wardrobes in the corridor, so in my second year I volunteered to live in the old wooden Army huts, where I was allowed a single room. There were five students in a hut with a bathroom at the end. I started teaching, thirty-eight 'backward children', at the age of nineteen. I enjoyed my career in teaching.

Cocoa and burnt toast at Edge Hill

Lois Ford 1946-1948

The Geography Group was a small one. We dealt with the geography of the British Isles in some depth, the geology, climatic conditions, land use, but I don't remember who taught it, it's nearly sixty years ago. The textbook we used was written by a Frenchman. The visit to Malham Tarn I remember distinctly, because it was so different from anything I had seen previously. Individually we had only visited places within a restricted area and books were our main source of information. Geography was taught in a room on the first floor in the main building. It was furnished with long tables and deep cupboards for storing maps. Library facilities were very meagre. I chose the Windermere Valley as the subject for my thesis, and produced a number of large-scale maps of the lake and its surroundings to illustrate different aspects of its geography together with an accompanying written account. Students tended to use their own home localities, for example those living in the mill towns would be likely to study textiles.

Professional Course

Principles and Practice of Teaching

Hygiene

Physical Training

Main Subject Geography

Courses at Ordinary Standard

English

Social Studies

Art

Needlework and Handiwork (Weaving)

I had taken Geography as a main subject in the Higher School Certificate, so it was a natural progression, though in spite of taking Geography as a main subject, it could only occupy a comparatively small section of the time available. A glance at my Teaching Certificate (see table above) indicates that the subjects covered were as shown.

The main objective was to prepare students to teach a wide range of subjects in the state Primaries and the then Secondary Modern Schools. This I feel they achieved. There were few wasted moments, and breaks during the day were for meals with some (not many) free periods. An Evening Meal was provided at 6.00pm, then we were expected to study in our rooms from 7.00 until 9.00pm, when urns of National Milk Cocoa were trundled round to the Halls, and the smell of burnt toast would permeate the corridors. There was a roll call at 9.45pm, and lights were supposed to be out at 10.00pm. The Halls of Residence were quite separate communities, within which we had 'families'. Our Second Years had 'daughters' in the First Year. It certainly helped to integrate students into College life. I was a member of a closely-knit group of eight. Each hall had its own Warden who had accommodation on the ground floor.

Our teenage years had coincided with the war years. Our wardrobes consisted of school uniform and little else. We were granted additional clothing coupons for items needed for College. The College blazer was worn with pride (Fig 5.4). My mother made me a housecoat and an outdoor coat, while I made a pair of slippers and I knitted a sleeveless pullover with grey welts and bands of different colours, using odd balls of wool. It stood me in good stead. The New Look came in during 1947 to 1948, and I remember a visiting former student wearing a long skirted garment in the latest fashion, and the feeling of envy I experienced. Our parents paid a fee, according to their incomes, but there was no grant for personal use. Most of us had little cash to handle. Few parents had cars, so it was necessary to use public transport. Our belongings, including sheets and towels, were packed into trunks and then transported by rail from door to door for a few shillings. Except at half terms we were expected to be on the premises at weekends. I remember one half term asking if I could leave earlier in order to catch an earlier connection at Preston: this was refused. I didn't find the strict regime particularly irksome as I had had a strict upbringing.

Food was still rationed. At breakfast two slices of bread were provided for each student. We were issued with a jar of jam or marmalade and it was replenished every eight weeks. Cartons of chocolate spread did not require points so they were used to supplement, otherwise the pots of marmalade went down to the breakfast table, day by day. An occasional treat to supplement the diet was the purchase of a chocolate cake from an Ormskirk confectioner. Saturday mornings were a time utilized for cleaning our rooms, so smaller items of furniture would appear in the corridors. We shared rooms, so there wasn't a lot of space for personal possessions. On Sundays we were placed on a rota to deal with the dishwashers in the kitchen to enable the kitchen staff to have shorter hours. A group of us had our cycles with us, so there were occasional rides to Southport or Formby. Some weekends I visited a cousin in Liverpool, but my parents never visited the College. Very few of the students had radios. There were no typewriters, so all the essays were hand written. For our map work we had spidery map pens, watercolours and Indian ink. When the finals were over short courses were provided in areas of the curriculum we may not have covered. I chose to do bookbinding, hence the cover of *Geology in the Service of Man* (Fig 5.3). The book is very thin, as post war paper was very poor quality. All students, first and second years, left at the end of the Summer Term on the same day.

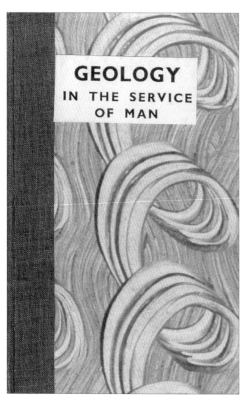

Fig. 5.3 *Geology in the Service of Man* (cover)

Source: Lois Ford

School practices were of course of paramount importance. My first practice, in a Secondary Modern School in Bootle, was a cultural shock. During Observation Day I witnessed boys having their palms slapped with a cane on forty occasions. We of course were expected to maintain discipline without its use. My room was a prefabricated building with bomb craters on both sides. From 11.50am onwards the room was used as a dining room, so we had to vacate it. My second practice in the autumn of 1947 was in a church school in Wigan with a class of fifty-five seven year olds. The desks of the old fashioned dual type were so close together one could only sidle sideways between them. After leaving College my first post was in an all age school catering for five to fourteen year olds, in a Victorian building containing six classrooms. There was no hall, no central heating, no telephone, no radio, no clerical assistance, no hot water, no staff room and the toilets were outside. Classes needed to contain more than one age group. At the end of each term some fourteen year olds would leave, and some five year olds admitted so there would the movement of some children from class to class. Heating consisted of four open fires and two stoves.

Every morning the woman caretaker lit the fires and carried in the buckets of coal for the day's use. For this she was paid a pittance. Fortunately conditions changed, and modern schools were built in the town as the population doubled in size. I taught until I was sixty-five years old, and consider that I had a very satisfying career. I have since attended a Keele University course in History where the oldest class member was ninety, proving that education is for life.

Lois Ford's copy of the 1946-1948 Edge Hill College Regulations identify, at Ordinary Level, a three–part syllabus of world geography, British Isles and use of maps, showing a continued focus on regional geography and on natural regions, with 'human response' expressed in the masculine phrasing of 'the study of *man* in *his* environment'. The Advanced Geography course focussed on British Isles and Western Europe, with a choice of systematic options (See Appendix V). Examinations were normally three hours, except when a thesis had been studied. In her 'Principles of Teaching' June 1948 exam Lois Ford had to answer five questions out of thirty-seven, in nine sections, three of which are shown in Appendix V. The two questions on geography ask about the use of maps and fieldwork in geography teaching, while other questions reflect assumptions of the time; that artists were men and that girls should be taught dressmaking and home furnishing.

Quite a community in Block A

Olive McComb 1947-1949

I didn't take geography at Edge Hill, but do remember 'Block A', where I lived for two years. I was told it had been the nurses' sewing room, for their relaxation during the war. I was also told that the grant to College from Lancashire for the new building was under guarantee that if there was a war it would be used as a hospital, as it had been expensive to build. I remember a two-page summary of coming back to Edge Hill after the war, which recounted stories of excavating the pit by the Ruff Lane entrance to dispose of amputated limbs[5].

---

[5] Betty Underwood considers this story most unlikely

Ruth Whitaker, Barbara Wilkinson and Joan Greenhalgh wearing their College blazers in 1949

Students in the cold winter of 1950-1951

The view from the top floor of Stanley Hall in 1951

'Three generations' Kath Cunliffe with her College 'mother' and her 'mother's mother'

Fig 5.4 Post war Edge Hill

Source: Kath Twentyman

Block A had about twenty-two student rooms and two staff rooms for two resident tutors, who each had a sitting room and bedroom. Miss Davies, Home Economics had rooms on the right, and Mrs Burdett, RE, very maternal and believed to be a widow with six children, on the left. When the Gardening tutor grafted a new rose which was accepted by the Royal Horticultural Society, it was named Mrs Burdett. Miss Davies was young, but died in her forties of cancer. A general Purposes Room was opposite, with a toilet. Bathrooms were round the corner.

Our rooms were in the two wings, but we were nominally allocated to Halls in the main block. I had a single room, very Spartan. Unlike the Main Block it was not furnished with Waring and Gillow furniture, but with a 'rag tag and bobtail' of ex War Department stock. There was an old hospital bed, a little folding card table, on which we played the oiuja board which took the varnish off and a utility wardrobe-cum-food container, though we had little food as things were rationed. We could spend up to two bread units per week at cafes in Ormskirk; this bought four currant buns. The walls were painted cream, curtains were cream calico, and I had a rug, which we used to beat in the corridor to get rid of dust. We cleaned our own rooms, and were on a rota for sweeping the corridors; cleaners did the bathrooms and toilets. Block A was quite a community; the huts were lovely. 'Tick in time' was 'ten to ten'. Reps went round to check that everyone was in their own room. Lights out was supposed to be 10.30. I spent two very happy years in Block A with a lovely view of the old water tower, like a Grecian temple or folly, to greet me each morning.

We went cycling and ranged far and wide over the Lancashire countryside. Now and then we had a glass of cider, though it was not a drinking culture; we didn't have enough money. I remember on one occasion giggling as we pushed our bikes up Ruff Lane, and one of the posh residents said she would report us to 'the headmistress'. We fell about laughing but heard nothing more. We sometimes went to the Hayfield pub, and as a treat had a poached egg on toast for one shilling and sixpence. Eggs were rationed, one per week. We once turned up to a College meal of mashed potatoes with green slime (spinach) with one poached egg in the middle. There was such a wastage of precious eggs. Teams of six to eight students were in the washing up rota at weekends. As a gardening student I made great thick salad sandwiches with two slices of bread. A few students who were not eighteen had different coloured ration books, and were entitled to one banana a week, though they never saw these. However one day the staff on the high table had banana split for pudding. At formal lunch every weekday we sat at different tables with a member of staff on each, and had to make polite conversation[6].

1950s Geography

In the early 1950s geographers made great contributions to planning for post war Britain. Some geographers were pushing to change counties, which they argued were outdated, as their boundaries had not changed with city growth. They advocated larger regions based on divisions used in the war. E W Gilbert of Oxford University considered the eleven Civil Defence Regions to be more effective for

---

[6] Olive McComb (née Parr) (1947-1949) Questionnaire response and Telephone interview 22 December 2004

planning purposes, and believed that regionalism would be stimulated by the organisation of broadcasting into Midlands, Northern or Welsh regions. However geography fieldwork had not advanced by very much; field equipment was dominated by the needs of map making. Putnam in Taylor (1951) advocated a notebook, coloured pencils (in packs of twelve or twenty-four) compass, plane table, soil auger, camera and tripod, together with a car or jeep, though in 'Lab. Studies' of maps he advocated, in addition to dot and isopleth maps, 'recent' aerial photographs.

Post war publications of the Institute of British Geographers were mostly on either geomorphology or economic geography, with biogeography and medical geography recognised as new fields. School textbooks in the 1950s, which now had more illustrations and activity work, helped geography to remain a real world study by sample studies and case studies (Walford 1995). Geography texts in the 1950s had amazing longevity. Philip Lake's standard book on Physical Geography, first published in 1915, was last reprinted in 1958, and was used for many years after that for sixth form and first year University work (Steel 1981:138).

Geography at Edge Hill In the 1950s

In 1950 Miss D M Tovey joined the staff at Edge Hill as lecturer in Geography in January, while in September 'we hope to welcome Miss G A Williams and Mrs Muir'[7]. In 1952 Miss Tovey is mentioned as lecturer in geography (Fig 5.5).

> 'Second Year Students returned to College in September 1951 in blazing sunshine and found that the top floor rooms of each wing had been redecorated in peach, pale cream and pale green in place of the dreary post war khaki… Geographers and Biologists went on a whole day expedition to Silverdale'. (Edge Hill College Magazine No 61 1952)

After twenty years at Ormskirk, the Edge Hill College Magazine of 1953, noted that 'two more of the huts have been transformed, one to a music room and the other for the exclusive use of those preparing to teach infants'. This latter may have been Block A, though several students remember living from 1958 to 1960 in what is now the geography building. Reports in the magazine continue.

---

[7] Edge Hill College Magazine No 59 1950

'We may live in fairly rural surroundings but we are not isolated from the world of today... saw a Gaumont British film on the British monarchy... A horse chestnut was planted on the West Lawn and there was a large bonfire'.

Fig 5.5 Miss Tovey (left) and Mrs Muir: Edge Hill Geographers in the 1950s

Source: Kath Twentyman

In the 1957 College Magazine is the first mention of Anna Cooper, later to be Head of Geography, with a Book Review on *Geography in Schools,* a textbook, written by Grace Wood, and published by Blackie, at nine shillings and sixpence. It was a practical book for primary schoolteachers; following a brief initial statement about the 'spirit and purpose of geography', it dealt first with map work, integrated with local geography. In 1959 Miss Bain, the Principal, commented that she was looking forward to new buildings to house the increased student numbers, and that numbers of staff were also gradually increasing. A film on the conquest of the atom, produced by the National Foundation for Visual Arts, was presented. The College magazines continued to feature the annual **'In the Grounds'** accounts, noting further gifts of trees and plants, and that flowers were grown for cutting. 1950-1951 was a hard winter (Fig 5.4) in 1952 there were problems of weed control and a course on herbs was introduced in 1953. Some further extracts are shown below (Fig 5.6).

1950: The Staff Garden is enriched with a strawberry tree, a Judas tree, flowering currants, a weeping cherry, flowering crabs... The unsightly traces of military occupation behind John Dalton Hall have at last been largely removed and a new flagged path leads across the re-sown grass towards the Baths and B Block. In front of Block A a riotously–coloured flower border is set off by the white wall behind... The First Year Gardening Students have cropped a vegetable plot, and the produce is being used by the Housecraft Department...

1951: ...A long, dull winter, and a slow, cold spring. ...All through the drab months... there had been splashes of colour here and there in the rock garden... Our Student-

Gardener delighted us throughout the year with her exquisite flower arrangements... Old students owe her a special debt for her care of the rock garden.

1952: The spring flowers and fruit blossom welcomed us back to College after the Easter Vacation... the soft fruits ripened earlier than usual and we were able to have black currant tart at the Going Down Party. Weed control is always a problem in such large grounds and in the early summer dandelions were so numerous that they provoked much comment. The small rose garden in the School Garden is now planted with thirty-six different bush roses... we are building up a collection of old fashioned and shrub roses. Flower decoration in the building has always been a feature of Edge Hill and we are continually experimenting with different kinds of flowers for cutting.

Fig 5.6 The College Rock Garden around 1950

Source: Edge Hill Archives

1953: ...It seems a shame we must leave College when the grounds are in their full summer glory... We have been in charge of selling of flowers and have found it a very flourishing business. A beech hedge has been planted near the pond... The most notable addition to the grounds was a young chestnut tree planted to commemorate the

Coronation. It is situated on the lawn outside John Dalton Hall.

This report was by Sheila Douglas who studied gardening as a main subject at Edge Hill from 1951 to 1953). In 1954 the report was compiled by Mary Coles, the first Gardening student 1933-1934, and Lecturer in Gardening from 1949[8]. She reflected on twenty-one years of the grounds development:

> **1956**: The Rose Garden has over fifty different varieties of rose.

> **1957**: The mild winter made it possible to have many classes out-of-doors.

Edge Hill College Magazines also recorded details of geography field trips, this time to Wales and showing clear signs of the survival of a deterministic approach to geography, see the italics below.

> **1959:** 45 geographers spent a weekend in Merionethshire studying the type of economic development *determined* by Highland conditions in West Britain… group work allowed for a detailed examination of the relief and structure of the Cader Idris area and for visits and discussions with local specialists on hill farming, afforestation, slate quarrying and hydroelectric power generation. Harlech Dome and Welsh Black are no longer a jumble of words'. (Edge Hill College Magazine No 68 1959)

Ann Bowden (1955-1957) remembers being taught geology and South America in geography by two lecturers whose names she can't remember. Students visited Parbold and the water tower to see Bunter Sandstone. Geography was taught in the main College buildings, and she believed Block A was used for RE lectures[9].

A happy band living in 'Block A'

Dorothy Collings and Margaret Catterall, 1958-1960

Block A had about twenty two girls, first and second years, under the watchful eye of Miss Redfearn, who taught handicrafts, and 'Tottie' Holmes[10], who taught Biology; they had rooms either side of the entrance hall, and 'ticked-in' students by 10.30pm (Fig 5.7). We lived a

---

[8] Later to become Head of Rural Studies
[9] Questionnaire response
[10] A nickname implying short, though called Totty Hemes by Kath Twentyman, 1950-52

most Spartan life. I recall arriving at College on my first day having walked from Ormskirk bus station with one suitcase, a bag and a biscuit tin containing one large cake. We each had a small room, containing a bed, dressing table, with drawers and bookcase, though I cannot remember any other furniture. When we were sent the booklist prior to going to Edge Hill, we were given the measurements of the window with instructions to provided our own curtains and cushion covers in which to put the pillows during the day time. We stuck our posters on the blue gloss painted brickwork. The floor consisted of the original floorboards with a strip of carpet by the bed. Each day we were expected to clean our rooms using dustpan and brush. The ablutions were primitive. The five or six wash hand basins that drained into a trough were in a communal room. The three bathrooms had curtains instead of doors. There were no showers. The facilities for washing clothes were even more primitive. This was done in the General Purpose Room, where the large sink was so high that my friend had difficulty getting her arms into it. Wringing out was done by hand, then the clothes had to be put on a rack to dry. Another of my friends circumvented the whole process by sending her washing home.

All the meals were provided in the main dining room, except for Saturday tea. For this we were each given either one egg or a small tin of baked beans, together with bread and butter and something for afters. Despite the primitive facilities we were a happy band and saw advantages in not being in the main building.

I can't remember much about what we were taught in geography, though I know we worked very hard without the aid of photocopiers or computers. The teachers were Miss Cooper, Miss Williams and Mr Williams. We went out locally looking at the strata of the soils, and discussing about the water towers. We also stayed at a Youth Hostel in Dolgellau, Gwynedd, finding out about the surrounding area, and we climbed Cader Idris. During the first year summer vacation I had to do research for a study of the area of Preston where I lived. I spent most of the holidays photographing shops, schools, hospitals and different styles of houses. I also interviewed the managers of small industries to find out just what they manufactured and where their workers came from. Many hours were spent in the Record Office tracing historical maps, then completing them in Indian ink and poster paint. Also a long essay had to be completed entitled 'Discuss the features of the agriculture characteristic of closely settled areas within the Tropics. Illustrate your answer by more detailed reference to one area of South West Asia'. I still have these two pieces of work. They are not as professional looking as work done today, but a lot of effort went into their production.

Fig 5.7 Miss Hemes and Mrs Burdett (front row extreme left and extreme right) resident tutors in Block A in the 1950s

Source: Kath Twentyman

Conditions were quite hard for students on their post war return to Ormskirk. Staff expected Victorian strictness, with students in their rooms by 10.00pm. Geography teaching at Edge Hill, as in all training colleges, continued to be part of education, while several new Geography lecturers had Welsh connections. Even approaching 1960 life was Spartan for the last group of students resident in Block A. By the 1960s conditions began to change, as recounted next.

6: Back to the Future in the Sixties

The 1960s saw many changes, to society, in geography and at Edge Hill. The 1950s rationing had ceased and bombed cities were being reconstructed. Foreign travel was difficult because of currency shortages, but the GA began to organise study tours abroad, and the Liverpool branch regularly chartered steamers on the Mersey (Balchin

Fig 6.1 Edge Hill from the air in 1966. The Geography block can be seen in the middle foreground, with the Arts and Sciences block to its left, and Ethandune in the bottom left. Some wartime wooden huts are visible on the left, while behind the College are 1960s Halls of Residence. The running track is clearly seen.

Source: Edge Hill archives

1991). Jean Cowgill, a former student, remembers Edge Hill geographers in the early 1960s being dominated by the Welsh, while lecturers Rachel Bowles and Bill Marsden give their recollections of

Edge Hill geography later in the 1960s, and Sue Sumner recounts a romantic encounter.

Geography In the 1960s

The 1960 copy of the journal *Geography* carries an introduction by the retiring GA president, J A Steers; a highly influential physical geographer who wrote many books on the British coastline, including *The Sea Coast* in the New Naturalist series written for a wider audience. He was concerned about the development pressures he saw there and advocated coastal preservation, a view which was taken on board by the National Trust in their later Operation Neptune attempts to purchase valuable stretches of the coastline. His ideas were also responsible for the recognition of Heritage Coasts. In 1966 L D Stamp's 1929 textbook *The World* was in its eighteenth edition, which reflected changes in independent countries in Africa, and the change from the use of degrees Fahrenheit to degrees Centigrade for describing temperature, though the use of millimetres for rainfall was not included. In a 1968 IBG Memorial volume to L D Stamp some geographers expressed concerns about land use changes since the 1930s; R H Best on urban expansion, and S H Beaver about derelict land, especially the spoil heaps of South Lancashire. He also noted the expansion of new industries, like motor vehicles at Speke.

During the 1960s University geographers began to embrace the scientific method and quantification, in a 'new geography' aiming for objective proving of laws. Models, not concrete or actual, but theoretical expressions of the real world, began to use statistics in what geographers call the 'Quantitative Revolution'. New methods were applied to urban geography, aiming to describe the internal structure of cities (Johnstone 1979:64). R J Johnstone describes the late 1960s to early 1970s as fairly traumatic years for geography, as environmental concerns became stronger, and a shift in geographical topics occurred from supermarkets to poverty, Third World and inequalities (Johnstone 1979:147).

Economic changes were occurring in the country and a post war economic boom was beginning: planning took on a major role, with geographers presenting reports on future land use patterns (Johnstone 1979:25). Chorley and Haggett were writing texts in the 1960s introducing the new geography to teachers, though human geography still hoped to explain the world system of '*man* and *his* natural environment' (1967). There is however little discussion of philosophy and methods in geography apparent in Edge Hill student recollections of geography learned in the 1960s, unlike discussions occurring elsewhere in academic geography.

Geography at Edge Hill in the 1960s

The **1960s** saw many changes; the re-introduction of a three-year degree course[1], and the first male staff and students; in 1959 forty-two men students joined one hundred and twenty women[2]. The men students were referred by their surname, as 'Mr Smith', for example.

Fig 6.2 Miss Mary Coles, Head of Rural Studies, with the Grounds Staff

Source: Bernard Williams (left)

Stanley Hall became a male hall of residence, as some believe did Block A, which later became an education block. In 1961 the College acquired some extra land and a running track was laid. Provision was made for rugby and cricket, as seen on 1960s aerial photos of the campus (Fig 6.1). Football and rugby pitches for the male students were provided on the orchard and hockey pitch. The land in front of geography was laid out for hockey, with lacrosse on the other side of the main entrance[3]. Edge Hill had a strong Rural Studies department in the 1960s; huts for chickens can be seen on the aerial view. The

[1] Edge Hill had provided three-year degree courses for suitably academic female students before the turn of the century in its original location in Liverpool
[2] Edge Hill College Magazine No 68 1959
[3] Derek Sumner 22 July 2004

number of Edge Hill staff rose sharply from twenty-six in 1956 to nearly a hundred in 1966. In 1964 Mr Ken Millins, a former HMI became the first male Principal for Edge Hill; and Tom Eason agreed to write a history of Edge Hill College. Montgomery (1997) identifies this period of growth as causing pressure on accommodation and on tutors who had little secretarial support.

It was felt that colleges should train more graduate teachers; Edge Hill Training College was renamed Edge Hill College of Education, and a new four-year BEd degree from Liverpool University was offered. This change was also accompanied by a large increase in student numbers; by 1966 the College had 750 students, nearly triple the wartime number, and further new buildings were

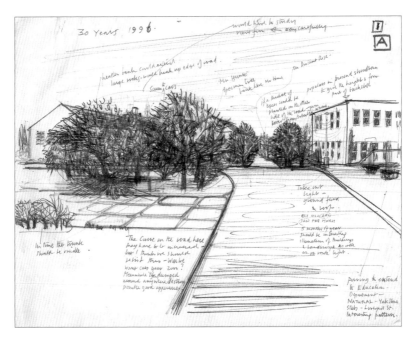

Fig 6.3 The geography block (left) in 1966 with sketches of proposed plantings, most of which were not carried out

Source: Edge Hill archives

planned. Also included were multi-purpose teaching blocks to replace the Army hospital huts, which in 1941 had been intended to last only for five years (Montgomery 1997:60).

The 1960 College Magazine noted that the Science and Art building 'was rising between Block A and Ethandune,'[4] which had been acquired as a women's hall. In this year of the second intake of men students, there are 'now 100 men and 200 women'. Most of the men were established in the Huts, which they called cabins. 'The crocuses under the lime trees are very fine.'[5] A B Dale (1961-1964) had a room in 'The White House' in his first year, on the left hand side of the back door in one of the larger rooms. He remembers visiting the beach at Ro Wen in Wales to help Mr Williams.

Following the construction of the new block built near Block A, a group of staff, including Miss Coles, Head of Rural Studies (Fig 6.2) and Edna Wright of the Art department, presented plans to the Principal Mr Millins, for the planting of trees and shrubs to soften the new buildings and link with the older ones. They presented some photographs of the Geography building taken in November 1966, together with some sketches of proposed planting for the spring of 1967 (Fig 6.3). Following advice from Ken Hulme[6], Director of Liverpool University Botanical Gardens at Ness on the Wirral, and like Edge Hill on New Red Sandstone, they proposed plants designed to create visual unity between old (geography building) and new buildings. They also suggested how the 'motor car can be got entirely off the campus,' with certain areas to become traffic free. Other proposals were to introduce wild life, flood lighting and to use water as a landscape material. Sketches also showed alternative plantings projected into the future, 1996 and 2001. Some of these such as planting on the grass verge in front of the main building were not carried out, and only some of the specimen trees and tall pines were planted outside the geography block. However their report was far thinking and comparisons with the present shows that many of their ideas were incorporated into the grounds. Jean Cowgill (1961-1964) who did the three-year Teaching Certificate remembers geography at Edge Hill being dominated by the Welsh.

'Cymru erratic' a little bit of Wales in Lancashire

Jean Cowgill 1961-1964

I guess I fell in love with geography at school. It may have been something to do with the geography master who was a very good

---

[4] A late 1920s house on St Helens Road, see Chapter 2
[5] Edge Hill College Magazine No 69 1960
[6] Ken Hulme had been involved in designing the College Arboretum, with trees donated from Ness Gardens

teacher as well as being 'dishy'. I really loved geography - the best subject in the world. I followed my aunt to College; she was at Edge Hill in Liverpool. In 1961 we did two main subjects – geography was a must; I took Physical Geography with Art, which linked well together. The Geography lecturers were all Welsh, even Miss Cooper, who was very precise[7]. The department was a little bit of Wales in Lancashire transplanted onto the Lancashire Plain, 'Cymru erratic'. I remember the melodious voices (sleep inducing) and the wonderful pronunciation of Cader Idris and Tal-y-llyn Mudstones. I enjoyed all the fieldtrips, except when I kicked over a hornet's nest on the Yorkshire Dales excursion. I surveyed Skelmersdale before the New Town and did a local study in the West Riding. I remember a meteorology task, when I taped a weather forecast for a week and did frenzied drawings of symbols from the television. I never enjoyed this subject. It would have been easier with a video or even access to the Internet (but these were not available then). Maps were reproduced using 'drawing tables'. After Edge Hill I taught geography in a secondary school, then moved to a field centre, teaching environmental studies. Now semi retired and living in the Lake District I have done all the geology and geomorphology courses available from Lancaster University

Fig 6.4 Anna Cooper, Head of Geography 1957-1982

Source: Edge Hill archives

In 1961 Miss Smith, Principal from 1920 to 1941 died; she had overseen the Edge Hill move from Liverpool to Ormskirk. In that year 'Geography staff took a coach tour of SW Lancashire for students to learn about the environment of the College'[8].

In September 1965 Edge Hill's prospectus lists four geography staff; Miss Anna Cooper, HOD (Fig 6.4) Miss Rachel A Hirst (now

---

[7] Note that Ann Smith, appointed 1962, was from Lancaster, and was not Welsh
[8] Edge Hill College Magazine No 70 1961

Bowles) Miss Ann Smith and Mr A I Williams. Anna Cooper had been Head of Geography from the 1950s and taught historical geography, as did Ann Smith who joined the Edge Hill geography department in 1962 from Penwortham Girls Grammar School. Rachel Hirst, appointed to teach on the new BEd degree courses, found Edge Hill geographers a happy team, who created new courses without the validation guidance of today. She delayed her departure from Edge Hill, because Anna Cooper went on sabbatical to Leicester. Mr Ivan Williams enjoyed taking students to the Welsh coast, at Aberystwyth and Borth, and students helped his research at Ro Wen, the sand and shingle ridge south of Barmouth in Cardigan Bay.

The Edge Hill prospectus for 1966 records that Mr Williams had left, for Didsbury College, and Mr Vic Keyte, Mr W E Marsden and Mr Geoff Richardson had been added to the staff complement. Vic Keyte appears as a charismatic geographer, described variously by colleagues as a biogeographer and historical geographer. Bill Marsden was responsible for geography education and also found Edge Hill staff 'nice people'. Edge Hill geography was very different then. The students were mixed, but there was only the BEd degree. Geography used the Porter's Lodge for some time. Geoff Richardson, recently returned from Australia, was a resident warden and may have lived in the lodge, as well as above the College bar[9]. Rachel Bowles and Bill Marsden remember their time at Edge Hill in the 1960s.

The start of a new era - recollections of a young lecturer

Rachel Bowles

In at the start

In 1965 the Colleges of Education were on the brink of developing new courses and required lecturers with both teaching and subject knowledge who would be able to develop the new BEd degree, a four year course to replace the two and three year Certificate courses. In the previous few years I had acquired a research degree in geomorphology, become a founder member of the British Geomorphological Research Group, was a committee member of the Geographical Field Group, had set up the geology teaching at The Latymer School, Edmonton, started an Explorers Club to ensure the children of Tottenham and Edmonton at least got out into Epping Forest (soon they were exploring Wales) and, to balance this, sang

---

[9] Information from Ken Harrison, formerly in Media Technology

with the school madrigal group and choir as well as choirs in London. Consequently when my attention was drawn to an advert in *The Observer* stating a lecturer was required in the Geography Department, Edge Hill College, to teach geomorphology and be able to be involved in field courses. A quick look at the map confirmed that this met my criterion of 'a teacher training college wanting fieldwork, near a university town with a decent choir' I put in an application.

The letter, on March 15th offering the post from P K C Millins included the sentences 'Do please now regard the College as your "home"…' and 'It may be sometime before you hear from the Authority…' The former request was never in doubt for all the staff had made the applicants welcome - and the wagtails on the front lawn emphasised the security generated by the College. The latter forecast was borne out by a letter, dated Oct 10th in which a minute stated Harry Foster (Education) and Miss R A Hirst (Geography) were approved as lecturers. We actually joined several other new lecturers, mostly men, but also in Art Mrs Di Davies; we two became the youngest lecturers in the College. At Latymer the news was received with mixed feelings as I was yet another member of staff who had defected to a training college (Strawberry Hill, Twickenham, and Brighton had taken Music and Science staff respectively).

Anna Cooper welcomed me immediately but suggested details should not be worked out until 'early next term' for two reasons 'Firstly Mr Williams may be in the USA… secondly, we are obsessed with exam marking and school - practice and major curriculum revision on a College basis.' This set the tone for my next thirty eight years in teacher training - one was always in the middle of an old course, starting a new course and planning the next new course - around changing staff supply. The exam marking and curriculum planning was not new; the last commitments at Latymer included planning meetings for devising the new CSE syllabi in which much time was spent agonising about teacher assessed course work - a very new concept at that time. Mr Williams did not go to the USA so we were four with Anna Cooper responsible for the economic side of Geography (in the Wilfred Smith tradition) and Ann Smith responsible for historical geography. Her appointment was specifically to build up this branch of the new degree.

Subject concepts first - then application

As in school the training colleges were subject led, i.e. one taught one's subject concepts and illustrated with field and research evidence and then taught how this information could be applied and skills developed... The new experience for me was going into the primary school. We taught five certificate education groups:

i) Secondary (11-15 yrs) in which Geography figured as 'any other subject' combined with Physical Education
ii) Junior/Secondary (9-13 yrs)
iii) Junior (7-11yrs)
iv) Infant/Junior (5-9yrs)
v) Nursery/Infant (3-7yrs)

The education groups remained the same for the Bachelor of Education Degree, a four-year course for a degree validated by Liverpool University. But the course involved two main subjects, which were studied to a higher level than in the certificate course and in the fourth year only one subject was taken, with Education. This was very much the pattern nationwide in most subjects and remained so for about a decade until the cuts begun by Margaret Thatcher around 1977. Thereafter the role of education in the degree began to assume dominance, eventually to smother any efforts to retain a pure academic subject approach in conjunction with school application. Subsequently academic geography developed separately with the current consequences of having primary and lower secondary teaching being in the hands of non-specialist teachers who are ignorant not only of geographical concepts and skills but also of the progression in these areas. To some extent this has also happened in all the 'foundation' subject areas that include those subjects often chosen in the past for joint study with geography.

Physical Education has had a tradition of being taught alongside geography (all those knee injuries required a second string) and it was the PE staff, in particular Misses Gardiner, Blanchard and Morgan who inducted me into recognising schools of excellence (they also chided me as I wolfed down tea before dashing into Liverpool for a Philharmonic choir rehearsal). One infant school near Preston provided a vivid admonishment from the head, 'remember, children are never too young to understand maps' as she directed me to mount the largest OS chart (equivalent to the current 1:500) for the school area on hardboard, protect with clear plastic and use as a floor map. Subsequent research (Spencer and Blades et al, 1987; Catling and others, 1979) has supported this contention. Alongside the practicalities of understanding school practice supervision new courses had to be written and field courses planned. It was a good time to be innovative; quantitative geography and problem solving were coming to the fore; systems analysis and detailed field measurement were moving from the sphere of research into the school curriculum and the increase in university places had yet to begin. We still had quality students with whom we could have philosophical conversations as well as introduce to new ideas.

Fieldwork and yet more fieldwork

Fieldwork played a considerable part in the curriculum. Mr Ivan Williams, though a generalist, was knowledgeable about the Welsh coasts, in particular the beach formations of Ro Wen at Barmouth. Subsequent Welsh field trips were based on Kirkby Brookfield school's field centre at Dolcorsllwyn Hall, Machynlleth, in the Dyfi valley where we explored the Borth formations, the market settlements and the glacial formations on and around Cader Idris. Whilst at the field centre we were entertained with accounts of the sociological effect of residential field weeks upon the overspill Liverpool estates. Though this was the early 60s there were still instances of children being sewn into their vests for the winter; there were also numerous instances of 'notorious' children succumbing to the calming effects of the centre and, through several visits, during which they had teaching in all the 'foundation' subjects, becoming responsible citizens. This was in the days before 'lowland' LEAs had begun to impose restrictions for health and safety, soon to be followed by all LEAs. Other field courses took us to Arnside, Lancaster and Doncaster. On one trip when we were studying glacial features about Langdale it was interesting to find the PE women opted for the 'A' team who took the longest, hardest route of investigation. The first year students were always inducted in fieldwork at Arnside – this had to take place before October 16 to be sure of having decent weather and daylight.

Edge Hill is fortunate in having to hand a diverse collection of geological and geographical landscapes. There was a regular 'geological' field trip led by the local regional geologist Iain Williamson who educated us about the Coal Measures and the hidden and exposed coal measures. Basil Hall of the Soil Survey gave us several local field trips. I clearly remember him providing instances of the sand deposits upon which the St Helen's glass industry depended. It was easy to go out to see local evidence of glacial drainage and coastal dunes and the peat-ridden marshes between. Through Ann Smith we became conversant with the historical development not only of SW Lancashire but also of the growth of Lancaster. Ann and Anna ensured we were all conversant with the vernacular architecture (VA) movement (See Brunskill 1981) emanating from Manchester University. Ann Smith had links with Brunskill who wanted to see whether his recording scheme would meet our needs in geography. The collaboration could be seen in Brunskills' later popular texts. Time was spent devising a simple building record form based on the VA documentation but which enabled the students to recognise, through not only date plaques but also building materials and structural detail, how Lancashire settlement was influenced by geology and the development of a rail and canal infrastructure. These forms were used successfully later on other field trips. With Ann an historical geography course was developed based upon Quentin Hughes' definitive book on Liverpool architecture[10], much use of historical maps and directories tracing the development e.g. of Everton. This was good training for developing a similar course in my next job and later in locality writing.

Vital teamwork

In September 1966 the department grew with the addition of an urban geographer with main responsibility for secondary geography education (Bill Marsden); a biogeographer (Vic Keyte) and a generalist (Geoff Richardson). Ivan Williams had moved on to Didsbury College to a senior management position and later to Eastbourne as Deputy Principal. We could now really present geography courses in depth. Bill Marsden mentions his Southport work that I remember not only as a tour de force but also as being an exemplar for working with a planning department. Indeed, through Anna Cooper we developed links with many local government organisations in an effort to highlight the geographical reality of establishment decisions. (She would be bemused today by the antics of our Deputy Prime Minister's Office). Vic Keyte was not only a keen biogeographer but also one of the first orienteers. Formby Nature Reserve became a national orienteering course, opened by Chris Brasher, and it was the enthusiasm generated with not only the PE students but also the geography students that encouraged me later to ensure other students were aware of the many benefits of orienteering. We were a good team; Anna Cooper ensured we worked together to a high standard of excellence in all spheres of teaching including skill development. Ann Smith and I ran a cartography course whilst the field courses ensured competence in measurement skills and the specialist courses developed associated skills. We became skilled at producing map extracts with the new, sole, office Xerox photocopier and colour

---

[10] Note that the second edition Hughes Q (1999) *Liverpool: City of Architecture.* The Bluecoat Press celebrates Liverpool's fine buildings in full colour

booklets on the staff room Banda machine. Yet we felt the students still needed competence in understanding the information on OS maps particularly the one-inch seventh series (replaced by the current 1:50,000). In 1968 we planned a series of seminars that I have never had the opportunity to replicate. Time was made for weekly planning two weeks ahead of the actual seminar. The students were given preparation tasks a week ahead of the actual seminar when a theme was followed through – geological evidence, historical development, biogeography, settlement and so on, based upon the Geographical Association's *British Landscape through Maps* series. Misconceptions were faced and map interpretation skills really worked through. We used the Doncaster sheet, as this was also preparation for the Year One Easter residential field trip to Doncaster and we also used the Merseyside booklet in a similar fashion. Other College innovations were the development of personal tutor groups that actually worked. On reflection this was probably because of perceptive timetabling – there was plenty space between lectures – something else that has become whittled away for numerous reasons.

In 1967 Anna Cooper had sufficient confidence in the team to take a sabbatical year, from 1967 to 1968, following W.G Hoskins' Development of the English Landscape MA course at Leicester University (this was before the BBC series). We were joined by John Friar and Mrs Rolfe whom, I must confess, I really never got to know, probably because my mind was on other things, not only singing three nights a week but also traipsing to and from London every other weekend where I was to move, on marriage in the summer of 1968, to Avery Hill College, to a similar post. Ann Smith was Acting Head of Department and fielded directives from above, usually from Miss Stanton (Deputy Principal, a geographer who worked in the history department) and Mr Johnston (Head of Education) – good training for later becoming Dean. Geography pursued its innovating course making sure fieldwork held a prime place in the timetable and that this was also pursued, if possible, during teaching practice. There was always good contact with both Liverpool and Manchester Universities through Richard Lawton and Geoffrey North as external examiners and Geographical Association (GA) Branch activities. Some sterling work went into providing a programme of meetings and field excursions for local geography teachers.

Ann Smith reminds me that we must not forget Mrs Fry, our map curator and technician. She gave devoted and very competent service particularly in building up the historical record collection which included not only early OS maps but also tithe maps, census enumeration schedules, directories (*Pigotts* comes to mind) as well as seeing our field weeks were well resourced. She and Ann are still in touch.

In 1967 Heinemann came looking for writers for their new map series – the first of the series was on Yorkshire maps and a parallel volume was envisaged for Lancashire and Cheshire. The rest of the team were involved in other matters e.g. Bill Marsden with Collins, so I took up the challenge. I found the material recently and remarked to my husband that I had done a lot of work to which came a heartfelt response – 'you did!' If you need information upon Lake Coniston and the historical development of its landscape (charcoal burning and copper mining) Skelmersdale New Town, - and fourteen other studies of places in Lancashire and Cheshire up to the early 1970s, I still have the gathered resources. This information went to informing my later work at Avery Hill but the series was abandoned at short notice. When I started a family it was with the thought that I would be part-time for two or three years. Educational cuts saw to a different scenario. As a mother who had been involved to some extent with nursery and primary schools I was generously allocated the primary courses when these became detached from academic geography, but still retained school supervision of geography, PE and design & technology students in secondary schools. I also became involved in courses upon Environmental Studies and a BA/BSc Geography degree - but alongside, not integrated, with geography in education. Additional requests were filled to teach on academic courses on other Thames Polytechnic, later University of Greenwich, campuses including the architectural foundation course where knowledge of the vernacular architecture of Lancashire and Cheshire – and the rest, was very useful.

At Avery Hill there was close working with history - for my colleague was also a Hoskins fan. Edge Hill contact was never far away. Eric Riley became Head of Education at Avery Hill and always kept me abreast of the Edge Hill academic gossip. I always have a long Christmas letter from Alan Cowell who conducted the Gilbert & Sullivan performances in which, as an alto, I was Mrs Buttercup (Pirates of Penzance). I have retained contact with both Bill Marsden and with Ann Smith. We were a team who were happy to share ideas: maybe this prompted Bill to ask me to edit the Primary Guidance series for the GA when he became publications editor. Later two series were written to fulfil National Curriculum resource requirements for Scholastic – the Heinemann episode had been good practice. By the 1990s family demands were fewer, writing increased and the Register of Research in Primary Geography was begun – and Research Assessment Exercise was demanding its pound of flesh. It is no coincidence that many of the schools in the English schools survey of children's ideas on locality came from the North West – the geographical concepts behind northern children's understandings were in place through experience and were a useful control for other urban and rural schools in the investigation.

Half a century of change

The mid sixties saw the development of many educational ideas that have since been eroded by political expedience. Today these are in process of reinstatement – see the current pronouncements upon 'out of school experiences'. Yet the reason for expanding the training colleges – providing sound education degrees with strong subject bases has long been gone. With the changes since the mid-seventies have come other disturbing trends. Always curious about people and places I have regularly asked each cohort to sign up their 'home-home' area i.e. the place they felt to have greatest affinity. In the mid-sixties in both Lancashire and London there were students from every one of the then fifty-two counties. By the mid-eighties the cuts in travel grants restricted the Avery Hill catchment area to the South and South East. In the twenty-first century, discounting foreign students, the catchment has shrunk to a journey time of less than an hour with concentrations upon the Lea Valley and the Thames estuary. This alone means we are dealing with students who are geographically inexperienced in their own lives. Moreover there is a worrying situation today that, with the divorce of academic and educational geography not only are the subjects not getting to meet and cross-fertilise but the age phases are ignorant of the understandings and skills in the adjacent phases. Where once there was a team maintaining a continuum in subject teaching from phase to phase with united objectives (observation, enquiry and understanding of the world from local to global) there are, in numerous places fragments of subject teaching and no understanding of a continuum. This may seem a harsh judgement upon current HEI practice. The present Edge Hill geography education staff are doing a fine job – but admit they do not see much of the academic geographers nor of practitioners in different phases. How different from the sixties.

References

Brunskill, R.W. (1981) *Traditional Buildings of Britain: An Introduction to Vernacular Architecture,* Faber (2004) Third Edition, Orion Publishing.

See http://www.orionbooks.co.uk/4539-0/author-Dr-R.W.-Brunskill.htm for full list

Blades, M., Spencer, C., Plester, B. & Desmond, K. (2004) 'Young children's recognition and representation of urban landscapes from aerial photographs and in toy play,' in G. Evans (Ed) *Human spatial memory: Remembering where*, Mahwah, New Jersey: Erlbaum. pp. 287-308

Catling, S. (1979) 'Maps and Cognitive Maps: The Young Child's Perception', *Geography,* 64 (4)

Hughes, Q (1964) *Seaport: Architecture & Townscape in Liverpool,* Lund Humphries.

Spencer, C.P. Blades, M. and Morsley, K. (1989) *The child in the physical environment,* Chichester: John Wiley. 302pp.

Spencer, C.P. & Blades, M. (eds.) (2002) 'Children and the Environment,' *Journal of Environmental Psychology,* 22, 7-220

Spencer, C. (2004) 'Aerial Photographs and Understanding Places' in Bowles, R. (Ed) *Place and Space*, Occasional Paper 4, London Register of Research in Primary Geography, p 77-80

Personal Recollections of Geographical Education at Edge Hill in the late 1960s

Bill Marsden

In arriving in 1966 at Edge Hill for interview for a post of Senior Lecturer in the Geography Department, I was immediately struck by the luxuriously large staff room, the prospect of an individual office[11], and more free period time than I had been used to in twelve years of teaching. I was anxious to move into teacher education, and had shortly before the Edge Hill advert appeared been interviewed for but not appointed to a geographical education post at the University of Manchester. I had no doubts about accepting the offer of one of the posts in the geography department at Ormskirk. At the time I was appointed continuing growth in the higher education sector was taking place. Like other teacher training colleges at the time, Edge Hill had not only developed apace but had been transformed from a female only to mixed institution. The expansion was predictably accompanied by the recruitment of additional staff, and Anna Cooper, as Head of Geography, was able to make three appointments to her Department in the persons of Vic Keyte, Geoff Richardson and

---

[11] The staff room is the Staff Common Room in the main building and the individual office was in the Geography Building, on the left hand side of the corridor to the right of the entrance

myself. We joined two established members of the Department, Ann Smith and Rachel Bowles.

On appointment at Edge Hill, my main teaching commitment was to be secondary geographical education, for which the subject department was responsible. In addition I recall, if somewhat hazily, that local and regional geography, with emphasis on fieldwork, was also a responsibility of a number of us, and it was part of the work with which I much enjoyed. I remember joining in excellent field trips organised by Ann Smith in Arnside and Whitby, both involving the strong historical geography dimension in which she was so expert. I retain no documentation from the time, except detail of an elaborate suburban and central business district shopping survey that I organised in Southport and which I think must have taken place with Edge Hill students. The local and fieldwork elements were also designed to fit in with the emergence of the Certificate of Secondary Education. Here I had currently been appointed to work as an examiner with one of the two North-west CSE examining bodies. The local dimension was also of special interest in that I had been invited to write a textbook for Cambridge University Press on *North-west England*. Apart from being involved with secondary geographical education and local and regional study I had to contribute further to the timetable and was landed with teaching North America. I fear preparation for this was very second hand, and not at all research-based. In fact personal research was not part of the job specification at that time.

My experience of the Geography Department was in most respects very positive. Anna Cooper was a quiet, undemonstrative leader, a person of, dare I say, traditional integrity and conviction, committed both to her subject and to her students. I enjoyed working with all my colleagues and still maintain contact with Rachel Bowles and Ann Smith. In the wider area of the College I also believe I had some peripheral responsibilities in the Education Department, though cannot recall the detail. At this stage primary specialists tended to be suspicious of those who had had secondary only experience. At the personal level I re-established contact with a school contemporary of mine, Harry Foster, a senior member of the Department who later undertook a Master's degree and then a PhD both of which I supervised at the University of Liverpool. He produced high quality studies of local educational history which led to publications in research journals. Since his retirement he has become a distinguished local historian, author of excellent books on various aspects of the growth of our home town, Southport.

In the College at large, however, I felt little sense of rapport between the subject department and the central administration, at least as experienced by a junior member of staff. If my memory serves me correctly, information was passed down either through the Head of Department, or at formal, management-controlled staff meetings, The Principal, P K C Millins, was clearly a shrewd politician, preoccupied, in advance of his time I would say, with public relations initiatives, in promoting his institution as a centre of excellence. Following my appointment at the University of Liverpool Department of Education in January 1970, I retained some small contact with Edge Hill. At the time its courses were validated by Liverpool University. But the shortly to retire Professor Tempest and Principal Millins clearly did not see eye to eye and Edge Hill's external accreditation was moved to the University of Lancaster. I was hired by Lancaster on more than one occasion as an external examiner and had more insights from this experience into the workings of the Principal than I had had as a member of staff.

Before the end of three years at Edge Hill I had recognised that I was more interested in geographical education and links between geography, history and education, than in teaching straight geography. In the late 1950s I had completed an MA in the History of Education and it was the connections between this area and geography that intrigued me. My subsequent researches were largely on the geographical component in the history of education, and in the curriculum history of geographical and environmental education. But these I would have had little opportunity to follow up at Edge Hill. I therefore seized on the chance when the geographical education post at the University of Liverpool was advertised. My appointment there was significantly helped by the fact that history of education expertise was being sought as well.

This brief reminiscence has inevitably been subjective and limited, but hopefully not too speculative. Looking back, for me the Edge Hill experience proved something of a career pivot. For example, the stimulus of Ann Smith's interest in the historical field and from her splendid collection of historical maps helped to give me preliminary ideas for research to come. At Edge Hill also I had enjoyed positive contacts with local schools, and as my first term at Liverpool was largely a teaching practice term, I eased into it with some confidence. The College also provided me with tentative insights into the primary sphere of education which were to bear some fruit in later years at Liverpool. In the broader picture therefore my time in Ormskirk was a kind of rite of passage between twelve years of school teaching and nearly thirty to follow in an academic department of a University.

By September 1965, when there were 725 students at Edge Hill,[12] Ken Harrison was appointed as the first audio-visual (AV) technician, based in Block A, which since 1961 had been the base for the education department. The 1960s saw the start of the AV service loans, and the building had a workshop and also housed the dark room and editing room used by the College photographer. Edge Hill was an early pioneer in using closed circuit television (CCTV) and Ken Harrison later set up the NOAA satellite link and weather station with Julia McMorrow. The College Observatory was located in the U-shaped area behind the building. Students took readings everyday in front of the building at a weather station looked after by Bob Slatter, a Biology lecturer. In the room to the left of the back entrance there was a planetarium, a large plastic dome mounted into the ceiling which could be lowered and children could sit inside and see stars and galaxies projected onto it[13]. It was not until 1967 that the then new Education building, next to the Arts and Science block, was ready, and Block A finally became the home of geography.

Romance in the Garden

Sue Sumner

I started as a resident student 'Garden Girl' at Edge Hill in 1968 when Mr Millins was Principal, and earned £1 a week and my keep. Every morning at 9.00am I had to report to Mary Coles, Head of Rural Studies for my tasks of the day. I had to prepare Rural Studies practicals, I ensured there were flower arrangements on the tables in the main corridor, plants in all the front windows, and a big flower arrangement in Hale Hall. I had to water them all each morning. Rural Studies students had their own garden plots and the College grew its own vegetables. There was a lot of work to be done in the grounds. One day whilst planting spring cabbage with the then head gardener Des Armstrong along came 'Adonis on a tractor', Derek Sumner. Later I asked him to accompany me to a dance, and was mortified when he refused, but he offered me a ride on his tractor, and the rest is history. We are still living happily ever after and both are still working at Edge Hill.

---

[12] 1966 Edge Hill College Prospectus
[13] E-mail from Ken Harrison, December 2004

7: The Days of Diversification

In the 1970s geography at Edge Hill began to look more modern. New research-active members of staff were appointed and numbers of both staff and students increased. Details of geography courses in re-validation documents and prospectuses give a flavour of the times, while, in a series of 'windows', former Edge Hill geographers Peter Cundill and Neil Immins reflect on teaching geography, and Colin Pickthall remembers geographers as contributing to his dramatic productions.

Seventies Geography

The context for geography education in the 1970s was changing. Government involvement in industrial location increased, triggered according to some by the 1970s oil price rises, when concerns were first voiced about over use of resources. Giant firms, the precursors of today's multi-national corporations, began to develop, while nearer home New Towns, like Skelmersdale, were built and metropolitan counties like Merseyside were created. The 1970s also saw metrication, with kilometres replacing miles, and millimetres replacing inches of rainfall (in the academic world). Debate continued about attitudes to geography, model building and its refashioning in the so-called 'Quantitative Revolution', when geographers aimed to discover universal laws about human behaviour. US geographers had sought rules, like the Rank Size Rule, a mathematical relationship between the sizes of cities, while some considered Central Place Theory[1] geography's finest product; it was certainly influential in schools.

In the 1970s human geographers in particular began to react against earlier attempts in the 'Quantitative Revolution' to develop geographical laws, focussing instead on spatial variations and welfare geography. Another new focus was leisure.

---

[1] Central Place Theory (CPT) developed by Walter Christaller, proposed a series of patterns and mathematical relationships between settlements of different sizes. The general idea was that a large service centre would sit centrally to its service area, and that six smaller settlements would surround it in a hexagonal pattern. CPT predicted the number of people who would travel to each service centre, and was used by Dutch geographers in their plans for the settlement of the newly drained polder lands of the former Zuyder Zee. The North East Polder for example has Emmeloord as its central settlement with smaller surrounding villages in a CPT like arrangement. This proved unpopular, as the Dutch disliked being deprived of their view of the sea; later polders have coastal settlements. Christaller's theory was based on inland south Germany

Michael Dower (1970) predicted a huge increase in free time, and promoted schemes for Country Parks, to protect National Parks from car using masses. He urged geographers to focus on the new wave of car borne visitors to the countryside. Some geographers reacted against regional geography which they considered as too superficial: systematic[2] specialisms were preferred. H C Darby was influential in the UK approach to historical geography, making a detailed study of past geographies, based on source materials (Johnstone 1979). Geography was though still male dominated: Ron Johnstone (1979:6) notes that '*he* (the geography academic) will have undertaken original research'. The debate about the extent to which those teaching in Universities should also be at the forefront of original research still rages strongly today.

Fig 7.1 Edge Hill from the air in 1974. The oval shape of the Rose Theatre is seen on the left with the wartime wooden huts in the top left. The campus now appears more wooded and there are a few cars.

Source: Edge Hill archives

Geography at Edge Hill

There were two major changes at Edge Hill in the 1970s: BA degree courses were offered and these were validated by Lancaster University. The former had been envisaged in the 1972 White Paper, while the latter followed local government reorganisation in 1974.

---

[2] Systematic geography focuses on issues, as opposed to the older regional approach, which looked at all aspects of the geography of a region

Liverpool became part of the new county of Merseyside, while Ormskirk and Edge Hill remained in the West Lancashire district of Lancashire. Edge Hill became an Associated College of the University of Lancaster to help the newly established university build up strength in teacher education: degree ceremonies still in 2005 are held at Lancaster University[3]. In 1974, three major BA (Bachelor of Arts) degrees, Geography, English and Applied Social Sciences, were offered, so although geography has been taught at Edge Hill since its inception in 1885, BA degrees in geography started in the 1970s. The 1974 prospectus had a (rather dark) black and white photo display of 'starting points for work in geography', which illustrated streams, shopping centres and New Towns among other settings (Fig 7.1).

BA Geography at Edge Hill in 1974

Year 1
Regional Geography: Lancashire Studies, plus two options from Geology, Biogeography, Urban Geography, Industrial Archaeology, Archaeology and Statistical Methods
One systematic study from Geomorphology, Historical Geography, Urban Geography
5-day Field course

Year 2
British Isles: World Problems
5-day Field course

Year 3
The British Isles and World Problems

Year 4
BEd: Modern Geography

The **1974 Edge Hill Prospectus** described the BA geography degree as the study of *man in his* environment (my italics) still with a regional geography focus, starting in Year 1 with Lancashire Studies, and only one systematic option, from Geomorphology, Historical Geography or Urban Geography, in Years Two and Three the British Isles and World Problems, while BEd students took a fourth year course in Modern Geography. In September 1975 the first completely

---

[3] As Kelly (1991:440) put it in his history of Liverpool University, in 1975 CF Mott and Edge Hill 'found it fitted in best with their plans for degree courses to seek accreditation from Lancaster University'

undergraduate entry of 1000 BA students enrolled at Edge Hill (Montgomery 1997:61) though it is not certain how many were geographers. The description for Geography (BA BEd Dip HE) in the **1975 Prospectus** reflects an updated approach. The focus was now said to be on environmental conservation with a problem solving approach directed at the personal education of the individual students.

BA Geography at Edge Hill in 1975

Year 1
Changing Countryside
One of Biogeography, Industrial Archaeology, Geology, Vernacular Architecture, Meteorology
Techniques of Investigation

Year 2
Urban Studies, Transport Studies, Geomorphology

Year 3
Conservation and Development
Applied Studies options: Historical Geography, Urban Geography, Quaternary Studies
Dissertation and Projects

> The Geography course seeks to demonstrate the contribution Geography can make to the identification and understanding of environmental problems and the complex social issues which they raise today. (Edge Hill 1975 prospectus)

The first year in 1975 was based around the 'heritage element' in the Changing Countryside, with one choice out of six, which included investigation techniques. In Year Two emphasis was on the development of the built environment, of contemporary urban problems and the study of physical problems, while Year Three was termed the Conservation and Development Course, with a choice of options in Applied Studies together with the Dissertation and Projects.

A 1977 prospectus detailed the new degree in Human Geography, which involved periods of professional practice. By 1979 the syllabus had been updated again: Year 1 had a landscape focus, both physical and human, with a wide range of options, five in Year Two and twelve in Year Three, available in Part II.

The 1970s saw geography diversified to develop new BA degree courses, and new staff appointed, Peter Cundill, David Halsall and Paul Gamble in 1973. Peter Cundill was appointed to teach geomorphology, while David Halsall was a human geographer with a long-standing interest in the geography of transport. He analysed successive government Transport White Papers for measures of accessibility both social and spatial, and conducted research in Glasgow, North Wales and the Netherlands. He later developed a particular interest in feminist geographies, in women's issues of safety and work in urban areas. Paul Gamble, who had worked in India, taught Afro-Asian studies and some geography. When Paul had a sabbatical year at the School of Oriental and Asian Studies, London, he was replaced by Julia Franklin, now McMorrow. In 1975, Ann Smith who had taught historical geography became the Dean of the Faculty of Humanities, and was responsible for the courses now validated by Lancaster University and not Liverpool as previously. All new courses came to her office and documents were laboriously produced on a manual typewriter, with much cutting and pasting. Copies of all reports were dutifully kept, as they were so difficult to produce[4].

BA Geography at Edge Hill in 1979

Year 1
Landscape Studies: Physical
Landscape Studies: Human
Investigation Techniques 1

Year 2
Process Studies and Environmental Problems
Resource Management (Hons. Major only)
Urbanisation, Planning and Land Use
Urban Landscape and Community (Hons. Major only)
Problems of Development and Conservation

Year 3
Coastal Zone Management
Regional Development in the E.E.C.
Urban Recreation Management
Heritage Landscapes – Analysis and Conservation
Tropical Agriculture – An Ecological Approach
Planning for Development: the experience of the Third World

---

[4] Information from Pauline Bankes, 2004, who as Pauline Martland, was Ann Smith's secretary for many years

When Peter Cundill left in 1976, to go to St Andrews to further his research, Julia McMorrow took over teaching his newly designed second year course in Process Studies and Environmental problems. Later Nigel Simons was appointed and taught human geography with focus on resource and coastal management. Andrew Francis was appointed in 1977 in an interview which John Cater describes in the next chapter. Andrew writes:

> 'perhaps the explanation for my appointment ahead of John Cater (and Steve Kenny, who at the time was a Lecturer in Geography at Liverpool Polytechnic and is now a Pro-Vice Chancellor here at JMU) was that I had previous teaching experience at Kingston Polytechnic and then had relevant professional experience working as a Senior Project Officer for the Australian Commonwealth Government's Department of Urban and Regional Development followed by part-time lecturing at Middlesex Polytechnic. The appointment at Edge Hill was for both Geography and the newly established Degree in Urban Policy and Race Relations'. (Andrew Francis, 15 March 2005)

Andrew Francis joined the 'old guard' and was one of the first 'new researchers', with fresh, different ideas. He found it easy to make friends in Edge Hill, but saw no promotion prospects so left for the (then) Liverpool Polytechnic and returned to his first love of Town Planning. Steve Suggitt was appointed in 1977, as lecturer in geomorphology, when there were eight staff in geography, and the geography building was shared with history, which had a teaching room, two offices and storeroom. During Steve's twenty-eight years at Edge Hill he has had many offices: he started in the main building, shared with Andrew Francis in The Hollies on St Helens Road (now Edge Hill Enterprises) then later moved into the geography building. When John Cater was appointed Steve and he shared a room in the old wartime huts, now the site of the CMIST building. He moved back into the geography building, occupying a series of rooms as internal spaces were reorganised. In this section Peter Cundill, Neil Immins and Colin Pickthall write their geographical recollections.

People, Pollen and Teaching in the 1970s

Peter Cundill

Since I started as a research student in the Geography Department at the University of Durham in September 1967, my research interests have concentrated on two main aspects of pollen analysis:

The study of past landscapes through the interpretation of fossil pollen concentrations contained in peat and lake sediments. Initially this research was centred around the analysis of hill peat on the North York Moors but later work examined various deposits in the Howgill Fells and south-west Lancashire. More recently lake sediments in Fife and other locations in Scotland (Aberdeenshire and Perthshire) have provided the main focus for this aspect of research.

Fig 7.2 Scarth Hill fieldwork in the 1970s. David Halsall is seen on the left

Source: Peter Cundill

The study of the relationship between modern (present day) pollen 'rain' and vegetation. Some work was carried out on the North York Moors, but the main development of the research has taken place in Fife with the setting up of a long-term project at the Morton Lochs National Nature Reserve and short-term studies of oil seed rape pollen dispersal in north-east Fife.

I was working as a tutor in physical geography at the Geography Department Liverpool University when I was appointed to a lectureship at Edge Hill in 1973. My appointment was to replace a geomorphology lecturer who was moving to north-east England to take up a post in a Polytechnic there (either Newcastle or Sunderland Polytechnic).

Although I was appointed as a lecturer to teach geomorphology, in terms of my research and previous teaching I was really a biogeographer. I joined a small team of five other staff, Anna

Cooper (Head of Department) Vic Keyte, David Halsall, Ann Smith and Paul Gamble. It was a very lively and progressive group who cooperated well with one another and assisted each other in the wide range of teaching activities taking place in the department. For my first two years at the College I taught geomorphology to first and second year students studying for the Certificate in Education and BEd courses but I was also involved in teaching physical geography on the Environmental Studies BEd programme and at first year level on the North-West England course. Practical and local fieldwork formed part of most of these courses (Fig 7.2). Although the courses were part of education degrees and certificates in education, they were straightforward academic courses, teaching prospective schoolteachers the fundamentals of geomorphology and other aspects of physical geography. It appears at first glance to be a heavy teaching programme but it was taught in blocks because the students were out of the College for long periods of time on teaching practice. For example, teaching practice in 1973 to 1974 took first years out of College 6-23 May, second years 21 February-27 March (half the Spring term) and third years 4 October-7 December (basically the whole of the Autumn Term). This structure also permitted us to take the second year students on a residential field trip during the summer term and we managed to spend something like four or five days in North Wales (based at Bangor Normal College of Education) in each of the three years I spent in Edge Hill

Although lecturing was the main purpose of my appointment, I was also involved in other College activities. In 1973 all teaching staff in the College took part in teaching practice supervision and, even though I had no teaching qualification, I was not excused from such duties. Over the next couple of years I made many visits to schools to sit in, observe and comment on the lessons taken by students. Most of these students were studying Geography as their main subject specialism. Initially I was visiting students allocated to secondary and middle schools and therefore teaching geography but latterly I dealt with students in Primary schools taking lessons on a wide variety of topics. This was a very interesting and enjoyable part of the job and I learnt a lot about classroom teaching. As the new BA courses were introduced I no longer took part in the supervision of this kind primarily because the students on the BA degrees were not involved in teaching practice.

The pattern of teaching altered substantially with the introduction of the new BA degree courses in 1975. There was obviously some overlap with the old BEd/Certificate structure as the new courses were phased in but teaching in the new courses was no longer interrupted by teaching practice. These new courses were planned during 1974 and early 1975 under the auspices of Lancaster University who had taken over the validation of degree and certificate courses from Liverpool University.

Fig 7.3 Formby survey in 1975

Source: Peter Cundill

I had responsibility for setting up the physical geography aspects of the new BA degree and I designed significant parts of the first and second year courses as well as the whole of a third year optional course. I taught the GE101 module, Landscape Studies I: Introduction to Studies in the Physical Landscape, which consisted of twenty-eight lectures with associated seminars and tutorials. There was a separate practical course, Investigation Techniques: Introduction to Studies in the Physical Environment, which involved surveying at Ashurst Beacon, Skelmersdale and Formby Point dunes, followed up by practical work, drawing diagrams and analysing soil and sand samples and this was taught in conjunction with Vic Keyte, the biogeographer in the department (Fig 7.3). This fieldwork and practical module was closely linked to the theoretical work explored in the lecture course. I was due to teach the bulk of the second year module, Process Studies and Environmental Problems, but this was curtailed by my departure to St. Andrews and my replacement, Julia Franklin took over this teaching. It is interesting to note that it was only in 1992 that St. Andrews was persuaded to introduce an Environmental Problems module! If I had stayed longer at Edge Hill I would have also taught my new optional third year module in Quaternary Studies (I ended up teaching a similar course as soon as I arrived at St. Andrews) and this would have enabled me to introduce much more material on palaeoecology/pollen analysis, my main research area. I would have also continued to take part in the residential field trip to north Wales which now became structured within a new third year Investigation Techniques module.

From what has already been described it can be seen that when I first arrived at Edge Hill it was still very much a teacher training college with strong local authority ties (Lancashire County Council as it was then). This was reflected in the existing Geography staff who were all trained school teachers training students in the art of teaching geography in schools. They were a dedicated and professional group who embraced the challenge of writing new BA degree courses. The geography department was very small (four teaching staff) but dramatically enlarged when I arrived because both Paul Gamble and David Halsall started at the same time as myself. David had some school teaching experience but had been a research student at Liverpool University, while I was there as a tutor, and he completed his PhD at Edge Hill while Paul had experience of working in India as well as school teaching. Although initially I was the only member of staff with a PhD, two others had higher degrees in addition to teaching qualifications. However, there were only two of us (David and myself) with continuing academic research interest. The Head of Department, Anna Cooper, was a quiet and unassuming person but had a sharp tongue if her new young colleagues moved out of line! However, she did appear to enjoy the lively vitality of the newcomers and in general I enjoyed being part of a team led by a caring and tolerant Head of Department.

College terms were quite long (three terms each of twelve weeks duration each year) and therefore vacations were shorter than those in universities at that time and many of the College staff regarded these vacations as holidays. While they may have improved their lectures and carried out some administrative duties most of them did not consider that vacations were an opportunity for research. The College at this time did not recognise research as a legitimate use of staff time and therefore they did not overtly support it and this is a major reason why I moved to St. Andrews. However, the College did provide funding to assist the publication of my research article in the *Transactions of the Institute of British Geographers* (see below) and I did get support to attend conferences so there was some encouragement to keep up to date with research and teaching. The situation did gradually change with the introduction of the new BA degree and the arrival of further staff in Geography, Nigel Simons and Julia Franklin, both of whom had research degree backgrounds.

The heavy lecture programme and the necessity to produce new courses for the BA programme (and ultimately lectures and practicals for this programme) restricted my research activities, but several projects did bear fruit in the three years I spent at the College. Most of these involved the writing up of laboratory work carried out earlier, from my PhD research on the North York Moors (while I was based in the Geography Department at Durham) and work on the Howgill Fells from my time at Liverpool. However, Edge Hill did posses a research quality microscope in the Biology Department suitable for examining and identifying pollen and I did manage to borrow coring equipment from my old department in Liverpool to carry out investigations of peat bogs. This allowed me to start a study of lowland peat bogs in south-west Lancashire. The coastal bogs of areas such as Downholland Moss had already been explored by Michael Tooley in a study of land and sea level changes over the past 13000 years but there were a number of raised/domed bogs just inland from the coastal area which had never been looked at. Two of these, Hoscar Moss and Holland Moss were initially examined as part of an undergraduate BEd dissertation and this demonstrated the potential of these two sites. However, laboratory analysis had to wait until I transferred to St. Andrews and therefore the work is attributed to my time here. The lowland bogs of Lancashire are such rarities now that I was also commissioned to write an unpublished report on Hoscar Moss for the Nature Conservancy Council (now English Nature) but again this was done whilst I was at St. Andrews.

The Geography Department in 1973 was housed in a series of huts which provided comfortable teaching and office accommodation but were rather quiet and isolated of an evening. They had the reputation of containing a ghost (a deceased patient from the hospital days) and although I never saw this ghost I did have one unnerving experience. One evening during the winter I used the darkroom in the department to develop some photographs and had left the corridor lights on. No one was in the building and no one (as far as I could hear – and it was very quiet) entered and left the locked building, but when I emerged from the dark room the corridor lights were out.

Because there was not enough staff accommodation in the huts my first office was a room in the College Lodge or Gate House which I shared with Paul Gamble. The room was bright and airy and very pleasant but we were there only for about a year before the College authorities decided that they wanted the lodge for other purposes and we were transferred to two converted linen cupboards (actually substantial rooms) on a staircase in the main College building overlooking the sports ground. This location was much noisier and not as pleasant although it did have splendid views over the south-west Lancashire farmland towards Scarth Hill water tower. This remained my office for the rest of my time in the College.

The Geography Department staff in the mid 1970s also participated in sport in the College. For example, Paul Gamble, Nigel Simons and I all played in the College staff cricket team. This was at a 'fun' rather than more serious level but we all enjoyed playing against other colleges and some Secondary School staff teams and we did have a reasonable amount of success.

Articles attributed to my time at Edge Hill:

Simmons, I.G. and Cundill, P.R. (1974) 'Late Quaternary vegetational history of the North York Moors, I. Pollen analyses of blanket peats.' *Journal of Biogeography*, 1, 159-169.
Simmons, I.G. and Cundill, P.R. (1974) 'Late Quaternary vegetational history of the North York Moors, II. Pollen analyses of landslip bogs.' *Journal of Biogeography*, 1, 253-261.
Simmons, I.G., Atherden, M.A., Cundill, P.R. and Jones, R.L. (1975) 'Inorganic layers in soligenous mires of the north Yorkshire Moors.' *Journal of Biogeography*, 2, 49-56.
Cundill, P.R. (1976) 'Late Flandrian vegetation and soils in Carlingill valley, Howgill Fells.' *Transactions of the Institute of British Geographers. New Series*, 1, 301-309.

Geography and Education in the 1970s

Neil Immins

I came to Edge Hill in January 1975, when I was appointed as a senior lecturer in geography. My responsibilities included teaching on the human geography courses as well as being in charge of all the curriculum geography courses at both secondary and primary level. At that time geography students were being prepared to teach in either secondary or primary schools, and several geography tutors helped to staff the appropriate courses. The BEd primary and secondary specialist courses proved to be very successful. It wasn't long before a PGCE course was introduced into College, and I was responsible for the secondary geography curriculum course as well as the slow learners' environmental studies/geography course.

In 1982 I was transferred to the education department as a senior lecturer in education, and the provision of curriculum geography courses at both BEd and PGCE level gradually became the full responsibility of the education department. However, geography maintained its distinctive contribution in the BEd degree – during the first three years of the course, curriculum geography was taught as part of Area three along with history, religious education and science. In the fourth year, students specialised in one of those four subjects, and geography proved to be a popular choice. I retired in 1996, although I continued to teach part-time for another couple of years.

When I first came to College, life was very relaxed. In those days the number of staff at College was of course much smaller, and nearly everyone had time to meet in the senior common room (SCR) for morning coffee and afternoon tea, provided by the catering department. The table in the SCR became notorious for the wit and conversation of those who met there. It was also erroneously thought to be a hotbed of intrigue, causing one director to remove the table altogether - it was soon replaced.

Staff had time to engage in social activities, culminating in the Christmas party, including such events as Morris dancing and mini-pantos, during one of which events yours truly managed to put his foot through the floor of the stage in Hale Hall. Fortunately Health and Safety had not yet been invented.

The staff cricket team also flourished, and it was long thought to be the case that one of the requirements for appointment to the geography department was a willingness (as opposed to an ability) to play cricket. In my time at College many of the department did indeed (unlike me) display a talent for the game. Several of the department also appeared in Colin Pickthall's Shakespearean productions, with varying degrees of success.

I suppose, from a professional point of view, my happiest moments were spent supervising students on teaching practice, especially when they were teaching geography. I never ceased to be amazed at the commitment and quality of Edge Hill students. With the passage of time, one of the downsides of going into schools was that of being mistaken for the student's father – fortunately, no one mistook me for a grandfather, as far as I know. The comments of the kids were always interesting, and one brief conversation I will always remember went as follows:

> Eight-year old: "What are you doing here mister?"

> Me: "I've just come in to watch your teacher, and give her a bit of help"

Eight-year old: "That sounds a good job mister, I think I'll do that when I grow up"

I shall always remember my days at Edge Hill with affection and I hope I played a full part in College life - at various times I was a member of the governing body, an elected member of the academic board, the chairman of the staff association and the chairman of the academic council. For one brief period I reached the dizzy height of acting head of the geography department. I participated in many College dramatic productions and will always remember playing the title role in Richard III.

With any job, however, it is the people that are the most important, and I will always remember with great affection all my friends that I made through the geography department.

Paul Gamble: an appreciation of his life

Paul came to Edge Hill to teach Third World Studies, later to become Afro Asian Studies, and also to help with curriculum geography, and I think he also taught some physical geography.

He was an enthusiastic and dedicated lecturer, with an excellent sense of humour. His lectures were always enthusiastically received, and he had a rather idiosyncratic delivery - on those occasions when he elaborated on his notes his gaze would move upwards, leaving students with the distinct impression that he had somehow jotted some notes on the ceiling beforehand. He was forthright, and was always prepared to express his views forcefully in department meetings, which made for some lively exchanges with Miss Cooper.

He brought a wealth of experience to the curriculum studies course, and always enjoyed supervising students on teaching practice.

He entered fully into the social life in College, and was one of the stars of the staff cricket team, being an excellent all-rounder – he was far more agile and fit than most of the rest of the team, and woe betide anyone who got in the way of his ferocious returns from cover point. He also made a brief incursion into acting, being persuaded to make his debut in one of Colin Pickthall's Shakespearean productions. However his acting skills were nowhere near as proficient as his lecturing skills, and he never really came to terms with the need to deliver lines in an intelligible fashion. He will perhaps best be remembered for poking his head through the curtained backdrop, and in full view of the audience asking 'Am I on yet?'

He had a great love of the countryside, and liked nothing better than walking the fells of the Lake District with a group of College friends - he kept us entertained us with stories of his previous career, and of course, as is the case with most geographers, could never be quite sure of which precise direction we should be following. Quite often in the pub at the end of the walk we would engage in such childish activities as making up limericks, each of the party providing a line in turn - Paul always reduced us to helpless hysterics with his complete inability to ever produce a line that rhymed with the previous one.

Paul brought a breath of fresh air into the Geography Department, and he is much missed.

Edge Hill Geographers and Drama

Colin Pickthall

As a lecturer in English then European Studies at Edge Hill from 1970 to 1992, I began to produce staff-student performances of Shakespeare's plays in 1976, with *Pericles; Prince of Tyre*. The sequence came to an end in 1990 with *Richard III*. In between came *Measure for Measure*, *The Tempest*, *As you like it*, *King Lear* and *A Winter's Tale*. The same gang did drama workshops on *All's Well; The Merchant of Venice; King John; Hamlet; Henry V; Othello*, *Macbeth* and *Cymbeline*.

To pursue this venture, which grew beyond any calculation, I set up a staff-student drama society. Students drifted in and some caught fire from what was going on. The ballast had to come from the staff and the response to my invitation was extraordinary. English lecturers and Drama lecturers we could have expected and they were hugely talented and supportive. But in addition the Deputy Principal, Alec Gresty, a classicist, expressed a faint interest and ended up being a majestic Prospero. A principal lecturer in Education, Richard Foster, turned out to be a veteran of Oxford University Dramatic Society and performed wonderfully as the Duke of Albany in *King Lear*.

But, as it turned out unexpectedly, we had an influx of talent from the Geography department. In the early days they were not involved, but once Neil Immins and Nigel Simons saw what was going on, they joined in with enthusiasm and Neil's experience in particular spread outwards from our College group into dramatic productions throughout the North West.

Neil was a fantastic mantis – all arms and legs – superb (none better) in comic roles, but devastating in tragic roles like the Fool in *King Lear* and as *Richard III* (the second longest part in Shakespeare in which he never missed a line).

To record that Neil played Trinculo in *The Tempest*, Lucio in *Measure for Measure*; the Fool in *King Lear*; *Richard III* and *Macbeth* does not do justice to the massive influence he had and the hilarious influence he exercised. I recall him, as Lucio, losing his lines – an unusual event for him. He stepped into the audience and asked a chap on the front row to borrow his text. That chap was Steve Ryan, who was so impressed that he came to Edge Hill as a student, was one of our finest undergraduates and is now a senior lecturer in English at Skelmersdale College, encouraging local students into Edge Hill.

I had to play the Duke in *Measure for Measure* when another member of staff had to stand down. I am no actor but knew the lines. Immo was merciless. He would – in front of an audience of several hundreds up-stage me and wink.

We turned up for a Sunday afternoon rehearsal of *Winter's Tale* in the College Theatre to find that a group had left their gear out in there. Neil and one or two other colleagues seized the opportunity to perform on the abandoned drums – for over an hour. Nothing I could say would stop it so I went for a smoke in the rock garden.

The last play I produced at Edge Hill was *Richard III*. Neil took on the pseudonymous role with enormous courage. He told me that while he was confident of learning the lines and the choreography, he felt that the complex emotions of the part would be beyond him. Precisely because I knew, initially, his comic abilities, I wanted to squeeze all the grand comedy out of the portrayal. Neil agreed, and he did it. It was an enormous achievement.

In that same production there was one of the most incompetent pieces of acting I have ever witnessed. Another Geography tutor (in fact Head of Afro-Asian studies) Paul Gamble, had watched our performances over the years and said he would like to see what it was like to be on stage. I told him that it was the loneliest of places on earth and that your hands grew exponentially in size, but he wished to go ahead. Paul worked ever so hard. He knew his few lines as the third gentleman and the third carrier, but he could never learn the movements involved in being on stage. He would turn upstage instead of downstage and thereby completely block actors seeking to circumnavigate *Henry VI's* corpse. Famously, in the Bosworth Field scenes, Paul could not bear the tension of waiting for his cue to appear. At a quiet moment, his head appeared around a black scene flat at the centre of the stage – looked directly at the audience and disappeared back like a tortoise into its shell. He told me afterwards that he had tried acting and had no intention of doing it again.

A few years after Paul's first (and last) appearance on stage he was dead. Neil, myself, Paul and Tom Chapman were walking over Tryfan when Paul remarked to me on the fact that his leg was stone cold while he was sitting on a very warm stone. He had a cancer in his spine. Some weeks later I was tipped off by the chief consultant in Ormskirk Hospital that Paul was approaching his end and I was able to get there and at least hold his hand, sitting alongside his brilliant wife, Jean, a few hours before he died.

Paul was a good and a loyal friend; of independent and determined mind. He was a lousy actor, but in all other respects a star.

Fig 7.4 Nigel Simons (left) and Steve Suggitt relaxing after 1970s fieldwork

Source: Steve Suggitt

Nigel Simons (Fig 7.4) has recently returned to Edge Hill as Associate Dean. In his days during the 1980s as a Geography tutor at Edge Hill, Nigel found time to participate in the staff-student drama enterprises. Like Paul Gamble, his experience was slim, but unlike Paul, Nigel had a gift for enunciation that means he could put Shakespeare's words plum into the back row without too much effort. His portrayal of the King of France in *King Lear* was definitive. I remember clearly his haughty delivery of the lines:

> Not all the Dukes of waterish Burgundy
>
> Can buy this unpriz'd, precious, maid of me.

I also recall that, when his role in *Lear* ended in Act 1, Nigel happily played soldiers, servants and sundries for the next four acts.

In *Measure for Measure* Nigel played the sacristan. It is a complex and confusing play to produce. I played the Duke and in the complicated middle of the play I got myself lost. Nigel was on stage. I whispered to him 'fetch me the head!' Without a flicker, Nigel walked off and came back with the severed head (a cabbage soaked in red ink inside a hessian bag.) By this time he and I had chance to jump half a scene and get the play back on course.

The four of us also formed, for many years, the stable centre of the Edge Hill Cricket Team. Only Paul Gamble was very competent – Neil, Nigel and I were enthusiastic hopefuls. They were great days and enjoyable days. I never felt so fulfilled, as when I was producing and directing my great good friends in a Shakespeare play, and I remain amazed at how deeply involved were my geography colleagues.

In the early 1990s Jill Grinstead took over the role of producer in the staff-student productions. She switched us from Shakespeare to Brecht and the result was two wonderful, electrifying productions of *The Threepenny Opera* and *Mahogany*. Neil Immins, Nigel Simons and I took part. It was a new and important venture for us because both were 'musicals'. We managed, however, to find singing voices, and in the case of Neil, his singing voice was more an advantage than the rest of his abilities.

In the relaxed 1970s geography lecturers had time to play cricket and to act in Shakespearean productions. Having diversified into a BA degree, geography at Edge Hill had modernised, with academic staff following research in pollen analysis and remote sensing; provision for education becoming separate.

8: The Thatcher Years: the 1980s

1980s Centenary and Changes

For Edge Hill the early 1980s were an insecure and uncertain time, overshadowed by constant fear of cuts, as Thatcherite policy caused colleges to compete in market place. A threat of merger with the then Preston Polytechnic, now University of Central Lancashire, did not occur, and Edge Hill celebrated its **centenary year** in 1985 under Harry Webster as Principal. Following the 1987 Government White Paper *A Framework for Expansion*, Edge Hill became a University College with 6500 students. The balance of Edge Hill's educational provision swung for the first time to humanities and social sciences, not education as previously: Geography was in the Faculty of Humanities. Edge Hill along with other Colleges became independent from local education authorities, becoming an incorporated independent financial organisation. Lancashire had been a benign ruler of Edge Hill; it was a great challenge to have to run the College independently. A new principal, Ruth Gee, appointed as a change agent, decided that every course should have an IT component (Montgomery 1997:68) and in 1985 Macintosh computers were introduced, and a major programme of technical infrastructure began, with refurbishment and upgrading of buildings, and the construction of new buildings such as the Learning Resource Centre.

Edge Hill also took over Woodlands Campus, Chorley, for in service work, providing courses for Lancashire teachers. In this section, on the practice of geography, John Cater writes his personal social geography, Gregg Paget reflects on his brief time teaching urban planning at Edge Hill, Mike Pearson reviews his experiences of leading teachers' geography courses at The Woodlands campus. Nigel Simons reflects on resource management, a continuing geographical theme at Edge Hill, and presents some thoughts on the real world relevance of geography, while Joan Swinhoe notes the work she did to keep Edge Hill geography running smoothly.

Inequalities

The Thatcher government's policies to strengthen market forces and raise government revenue; the trend to property ownership, weakening trade unions and privatisation, even of essential utilities such as water supply, led to high unemployment and concerns over Britain's divisions. The economic benefits of the enterprise culture were not evenly spread, but overly based in the newer industries of southern Britain. Geographers studied the widening north-south gap in Britain, between rich and poor, black and white, sick and healthy, employed and unemployed, in new variant of the older regional geography. The spatial patterns in housing prices, or voting behaviour became new topics of a structural social or welfare geography, which focussed on the inherent disadvantages in the way that society works. The freedom of market choice also led to a later development of individual personal geographies. Edge Hill geography in the 1980s reflected the long-standing division between geomorphology and human geography.

Edge Hill Geographers in the 1980s

Early in the 1980s Gregg Paget was appointed, and Anna Cooper retired, to be replaced by Derek Mottershead, a geomorphologist, as Head of Department, and Edge Hill's geography transition to a male dominated environment was complete. Apart from temporary appointments as staff had sabbatical years to finish or gain extra qualifications, there was some stability in geography. Edge Hill prospectuses for the early 1980s describe the 'Subject Area of Geography' as having ten staff and offering BA degrees similar to those of 1979, while a 1985 Director's report by Ann Smith detailed some geography staff activities, and listed fifty-one students studying geography (exceeded only by sixty in English).

Early 1980s Edge Hill Geography Staff (1980-1981 Prospectus)

Miss A.W. Cooper BA (Hons) Wales; MA Leicester; DipEd Wales
J C Cater BA (Hons) Wales
W P Gamble, MA Cantab; MSc London; PGCE Oxon
D A Halsall, BA (Hons) PhD Liverpool, FRGS
N R Immins, BA Sheffield; MEd Nottingham; DipEd; CertEd Sheffield
V C Keyte, MA Oxon
Mrs J M McMorrow, BSc (Hons) Liverpool
N F Simons, BA (Hons) London
R J Slatter, MA Cantab; MSc Salford; PGCE London; MI Biol
S C Suggitt, BSc (Hons) Liverpool

Glimpses of mid-1980s geography staff activities are seen in a Faculty report by Ann Smith in the Directors Report 1985-1986. John Cater co-authored an article on *Ethnic Residential Concentration and the Protected Market Hypothesis* in Social Forces, produced a chapter in a book on *Race and Racism (*Methuen) and was a book reviewer for the journals of *Transactions of the Institute of British Geographers* (IBG) and for *Progress in Human Geography*. He was on the committee of the IBG

Urban Study Group and was also a member of a working party on the Future of Geography and Higher Education. Paul Gamble produced an article on 'Image-building and the Tunisian Tourist Industry' in *Tunisian Affairs*. David Halsall lectured on railway history for Liverpool University School of Continuing Education and for the Geographical Association (GA). He provided book reviews for the GA and for the National Railway Museum, and was still serving on the committee of the IBG Transport Study Group and was a convenor of a symposium on the Historical Geography of Transport. Julia McMorrow contributed two papers to an International Symposium on Mapping from Modern Imagery, and was elected to the council of the Remote Sensing Society. Derek Mottershead produced an article on 'Patterns of spatial and temporal variation in bedrock weathering and by coastal salt spray: a five year record' for *International Geomorphology* and was on the national council for the GA.

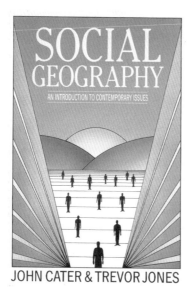

Fig 8.1 Textbooks written by Edge Hill staff in the 1980s

Geography students carried out a shopping survey on behalf of St Helens Borough Council, and in 1988 conducted surveys of minibuses in Ormskirk and Aughton. In 1988 a new BSc Geography programme began and led to the appointment of Ann Chapman and Kathryn Coffey as technicians. Documentation for this new degree provides more background into publications, consultancies and professional activities of geography staff. Six microcomputers were installed in 'computer lab', while Derek Mottershead argued strongly for more resources for geography.

Derek Mottershead[1]

When Derek was appointed as Head of Geography at Edge Hill in 1982, to replace Anna Cooper, he had co-written an undergraduate text, *Environmental Systems* (Fig 8.1) and wrote the second edition during his time at Edge Hill. He produced many publications, was a referee for the academic journals, *Area, Boreas, Proceedings of the Geologist's Association* and *Teaching Geography,* was an external examiner and Chair of the Heads of Geography in Universities in the UK. Derek undertook much research, particularly in the area of salt spray rock weathering of rocks, and initiated fieldwork in Mallorca, working on solution rates of limestone in the mountains around Lluc. He organised improvements to the fabric of the geography building, setting up and equipping computing rooms, the laboratories and lecture theatre, and strengthening the roles of technicians and cartography. From 1994 to 1996 Derek took over responsibility for preparing the Edge Hill 1996 Research Assessment Exercise (RAE) and was given the title of Professor. He gave his inaugural professorial lecture, 'The Wonderful World of Geomorphology' in November 1996.

Since leaving Edge Hill Derek has worked for Manchester Metropolitan, Liverpool John Moores, and Portsmouth Universities, and continues to collaborate with research colleagues in other Universities with good research ratings. He manages the finances of a learned society of international standing, and has been appointed a QAA reviewer.

Some Publications while at Edge Hill

Mottershead, D N, Pye K (1994) 'Tafoni on coastal slopes, South Devon, UK.' *Earth Surface Processes* 19.6: 543-563
Mottershead, D N, White I, et al (1992) *Environmental Systems,* Nelson Thornes.
Mottershead, D N (1989) 'Rates and Patterns of bedrock denudation by coastal salt spray weathering - a 7-year record.' *Earth Surface Processes* 14. 5: 383-398

---

[1] Compiled by Sylvia Woodhead

Mottershead D N (1986) *Classic Landforms of the South Devon Coast,* The Geographical Association, Sheffield.

Mottershead, DN (1983) 'Rapid bedrock weathering by coastal salt spray, South Devon.' *Journal of Geological Society of London,* 140: 321-321

A Personal Geography

John Cater

Prologue

No watch. Watery sunlight says it's four-thirty in the morning and the end of my second night abroad. Five to a room in a remote Austrian farmhouse, but far more comfortable than fifty three dozing fitfully on a speeding coach. Early teens, the fourpence a week insurance policy cashed, a late place, temporary passport and a thousand miles from home.

But what was the rumble that awoke? A farmer's tractor, primed for an early start in the fields? A steamroller? Several steamrollers? And several voices. I slid from the bunk, edged back the thin white curtain and, beyond a field of brilliant yellow sunflowers and my first, equally bright, sight of oil seed rape, two ancient army tanks, a personnel carrier, a jeep, old motor bikes – one with a sidecar – a policeman on a pedal cycle, and half a dozen villagers, men of varying age, dressing as they hustled behind.

It's August 1968, east of Passau and a mile from the Czech border. Within an hour the farmhouse is awake. Those who didn't succumb to the tanks or the excited voices of pursuing villagers lost the battle for sleep to military aircraft. I'll never know if anyone swam the river to cross the border, whether the troops kept people in, helped people out or were mere bystanders. At breakfast we were corralled into a single room and told that, today, we were not going beyond the farmer's gate, though by evening a few escaped and walked the local roads. Not brave, we headed gently uphill, away from the water. Early the following day we were moved on, driven to Attersee, some two hours away.

So why did I become a geographer? Geography was an OK subject at school, but no more than that. Too many capes and bays, rivers to remember, which town produced what. But geographers took pupils on school trips, and though I knew little of the geopolitical significance of what was happening, I knew people who did. That day, confined to the farmhouse, we talked for hours. I can still recall intrigue, twinged with apprehension, as we learned more of the history, the politics and the geography of an unseen corner of my Europe.

But perhaps the intrigue had started earlier. Aged ten or eleven, my enthusiasm won the one-inch sheet, yellowed and full of pin holes, off the Primary classroom wall. That summer I walked the contour lines, followed the watercourses, checked the church towers and spires, disappearing for hours. No self-respecting family will allow a child to do this today but, the eldest of three, with sisters to fight and a fourth on the way, I moved in with that map.

A year or two ago, the Teacher Training Agency ran a successful advertising campaign, 'No one forgets a good teacher'. No one forgets a bad one either, and after two years biking from council estate to grammar school I couldn't speak a word of German or understand the purpose of a single Physics experiment. But Geography was taught well, and taught by real people. People with a personality as well as a subject. People who played sport, ran the staff pantomime, turned out at lunchtime, after school and on Saturday. People you could talk to and relate to. I often ask myself, 'the subject or the people'? The answer is both but, perhaps wrongly for a book of this type, I've always been at least as interested in the 'who' as the 'what'.

Geography wasn't the best GCE, but tales of A level algebra and calculus and a Literature syllabus that went beyond Shakespeare, Hardy and ten twentieth century poets, coupled with a young teacher who was an inspiration and became a friend, kept me in the department. And, as the subject became more questioning and less descriptive, so my interest grew. Though a human geographer (a 'people geographer') at heart, the variety of the discipline, the interplay of science and social science, the literate and the numerate, fascinated me. But best of all, for a teenager unused to holidays, was the fieldwork and the freedom. Seven days on the Dorset downs, a week clambering the Lakes and the Dales, a spade broken on an ironpan, a flow meter washed to sea. Like it or not, I was too accident-prone for the physical geographers to retain.

Although there was no tradition of university education in my family, my street, my estate, filling Geography in the left-hand column of the UCCA form was easy. Visiting the places I'd nominated wasn't. Three of the six were in London – I know not why – but one trip to the Mile End Road to be interviewed by Alice Mutton made me wonder whether my geography was everybody's geography. And London didn't give me a sense of place, so half my choices were quickly jettisoned. At the suggestion of Gwyn Davies, the geography teacher to whom I owe a great deal, I 'threw in' as my sixth choice an outpost of the University of Wales, a new building, a new department and a new degree starting next September. Picking from six, I wouldn't have gone, picking from three, I had little choice.

Five days before Christmas, a bus to Northampton, a half hour walk, train to Rugby, train to Birmingham, train to Shrewsbury, two hour wait, the single track to Aberystwyth and thirty slow miles in Morgan's coaches. The journey could only be attempted on a weekday and took ten hours.

Two dozen hardy souls stayed for two nights and three days – free accommodation, free food, free entertainment (and long before the days of Aim Higher). But two things put a number one in the UCCA acceptance box. The first was the energy and enthusiasm of the staff and their engagement with us as geographers and as people. The second was from Llandeilo, very Welsh and, to me, very different. Come August I got the grades, she didn't, and a friendship crashed with the storm waves on Llangrannog beach.

Lampeter suited a small town boy two hundred miles from home. If you could hold a bat or pass a ball you played first team. If you showed the briefest inclination, you ran a club. If you happened to be in the Department on the first Friday afternoon, you ran the Geog. Soc. With friends, we booked bands with spatially illiterate agents one night and curious London academics the next. The intensity of the Cardiganshire's Welshness intrigued me, the unspoiled beauty of the country drew me – fortunate, given that every term was an unbroken twelve week block, the postal service was on three months strike and my parents didn't have a 'phone. A more self-sufficient eighteen year old returned home that Christmas, never to be a permanent fixture again.

The teaching was sound rather than spectacular, but the lecturers never felt their job ended at the classroom door. We sat on the floor in their rooms (a habit which lives with me to today) talked, questioned and learned. We ate in their houses, drank their wine and tried (unsuccessfully) to date their daughters. And, as ever, I remember the fieldwork. A week in Preston Montford, a village survey in West Felton, soil pits on the Longmynd (another broken spade) and a Dutch gap year girlfriend. That summer, I practised my transport geography, hitching the A5 to and from Shrewsbury.

By the time the final year came, I think I'd become a geographer. A human geographer for all but two weeks a year, when the call of hydrology at Teifi Pools or beach erosion on Cardigan Bay became irresistible. Two years of urban, social and economic geography and the opportunity to specialise away from impenetrable climatology and tedious pedology, and I was nearly hooked. I was approached about research studentships, but six years in one institution had left it toll on those senior to me, and the appeal of splendid isolation (and penury) was beginning to wane. I held Secondary PGCE Geography and PE places at Exeter University and Chelsea College, bizarrely located in Eastbourne and chosen simply on gender and sporting grounds (it was a newly co-educational women's PE college) and to one of them I expected to go.

There followed a summer of unease. A career was promised to accountancy and then banking; I turned up and reneged in a matter of days. I bought national newspapers, sent for application forms and, metaphorically at least, kicked the (non-existent) cat. Then, one Tuesday in late July, *The Guardian* advertised a Research Assistantship in a new polytechnic, a post for a social scientist to study some aspect of the South Asian diaspora in Britain. I applied, and with all other putative academics evidently out of the country, was called for interview in mid-August. The only fresh-faced young graduate in the room, perhaps I exuded desperation, because the 'phone call came the following day. Two weeks later I was sleeping on a floor in inner Liverpool, a city I'd never seen, reading social theory that I'd never understood.

I was Liverpool Polytechnic's only geography researcher and, in time, I think, their only PhD. A penchant for new departments had materialised again, and again I found supervisors and colleagues who treated me as part of a family. My research focussed on Asian residential and activity segregation in West Yorkshire, and Dave McEvoy, my first supervisor, loaned me a house in Bradford and a lifelong taste for heavily spiced food and Leeds Tetley's.

I also learned to teach. A year into research (and with only half a year's progress) I was a twenty-two year old Lecturer on a temporary full-time contract - economic geography for HND Business Studies to mature employees at night, cartography to women returners at Ethel Wormald on Wednesday afternoons, six first year tutorial groups, four statistics classes (I learn it tonight, you do it tomorrow) a final year Geography and Planning option, guest slots on others' courses and an interdisciplinary programme on Regionalism. A world expert I was not, and I've seldom worked harder nor across a more diverse range of topics.

The learning curve was a formative one, but twenty hours a week of all new teaching was hardly compatible with a barely started doctorate, and the summer of 1976 saw a return to Bradford's streets.

My research entailed a detailed survey of every inner city business, followed by the painstaking creation of residential grids from Census data and electoral rolls to get a sense of the relationship between residential and activity segregation. At the height of the Yorkshire Ripper scare, this was not a great time to walk the inner city's streets. The database, collected from paper sources and transcribed longhand, punched on computer cards and carried to the Polytechnic's only mainframe, took a lifetime. And the computer crashed halfway through its run every night, as my scant knowledge of Fortran and clumsy keying skills conspired.

Three-year research assistantships go quickly, particularly when we were concerned to have as high a profile as possible. I started as an academic author a week after graduating, combining some paid-for data collection on residential behaviour in South Wales with an article on the impact of superstores on cut-price petrol retailing – very basic distance decay stuff, but my first £40 cheque from a business journal. The following summer I was speaking at the first Geography and Higher Education Conference in Oxford, on matters not even remotely connected to my research, and playing an active role in the Institute for British Geographers' Young Researchers Forum. Spells on the Urban Geography Study Group and the Social Geography Study Group Committees followed, together with editorship of the UGSG's Research Register. Together we bid successfully for a sizeable Social Science Research Council grant to develop the work barely started in Bradford into Leicester and London, and I doubled my salary as the project's Research Fellow. Additional research work for the Institute for Environmental Studies and a fringe role in a major research initiative by the SSRC, the Small Business Research Unit, chiselled away at the months, as did a link with a newly emerging research team at the University of Oxford which led to a number of papers and chapters in the late 1970s and 1980s.

But 1979 arrived with an unfinished doctorate, research projects with barely time for completion, and a contract that ended on 31 March. Seventy job applications were submitted in January and February; the equivalent of seventy rejection letters came in. And then three calls for interview – a policy role in London, a three-year research post at Cambridge, and a permanent lecturing post in Lancashire. Three interviews and a flood of job offers. I took the first, rented a flat in Kingston and became a Policy and Corporate Planning Analyst. And then I took the third as well.

My Geography: My Edge Hill

A raw-boned twenty-two year old, I'd been to Edge Hill once. I sat in a dark Staff Common Room in the cold winter of 1976, shortlisted for a Lecturer in Geography post. We were seen one-by-one, me first, and then forced to wait. It had been my only interview for a full-time academic post, the panel was inscrutable and I didn't know how I'd done. Four hours after I'd left the interview room, the conversation as stilted as the SCR air, Andy Francis, back from Australia, was taken from us. Ten minutes later, leather footprints on the wooden floor, 'Mr. Francis has accepted the offer of the post. Thank you for your interest, you're free to go now'.

I didn't go back to Edge Hill, and I didn't expect to, my trips to Ormskirk being limited to annual cricketing defeats at Brook Lane. In the subconscious, perhaps I saw a place I'd be happy to work, a young team, a new degree, but interview defeats were like life ones – to learn from but not dwell on. But 1979 had a touch of desperation – seventy applications, seventy 'Dear John' replies. However, between application and interview I'd been offered and accepted the London job.

Should I withdraw? Edge Hill wanted me to come up on a Friday - I could swing that, and it meant a weekend with friends, travel expenses paid. Perhaps inevitably, the interview - my first and last all-women panel - was a breeze. I had a permanent job; now I had two. And I'd accepted them both. With some persuasion and an agreement to do some dissertation supervision in my own time, Edge Hill substituted September for an April start, and I headed back south.

Disgracefully, I hadn't decided that I would return. I'd simply bought myself time. Two things brought me back. My aversion to London, felt as a teenager, soon returned. The first floor 'studio' flat (box room) the crowds, the lack of personal identity held no appeal. And a policy unit before the May 1979 election was a different place to a policy unit after the 1979 election. Losing a car parking space for refusing to remove a Labour window sticker was trivial, but symptomatic. Being asked to write policy papers to reintroduce selective secondary education or close care homes wasn't. Was it the pull of Edge Hill or the push from the capital? Probably a bit of both, though perhaps the clinching factor was the prospect of creating a 'gap summer' travelling Interstate 95 from Boston to the Keys. I resigned.

I landed at Liverpool Airport, diverted by a baggage handlers' strike, on the third Sunday in September, and crashed on an old university mate's floor. Rucksack in hand, in slept-in jeans and unironed shirt, I fought my closing eyes through a one-hour Edge Hill induction the following morning, and then dashed off to find a home – arriving at a local New Town Development Corporation Housing Office just as it closed for lunch. On no one's list of expected arrivals (I didn't say, but this was hardly a surprise to me since I'd never applied) I nevertheless opened the door to a 'key worker' flat on the top floor of one of the more desperate housing developments in Skelmersdale twenty minutes later.

Before flying to the States, I'd spent a couple of days at Edge Hill early in the Summer Term meeting my first dissertation students, catching German measles and causing considerable concern to two old friends with whom I'd stayed, particularly the newly expectant Mum. Fortunately the scans were fine, and a first godson became six foot three of twenty-something beefcake. I also found my way around the 1940s hospital ward that acted as the Department, helped by the fact that colleagues like Nigel Simons and David Halsall were well known to me through the Institute of British Geographers and Geography in Higher Education networks. 1940s brick and whitewash was, however, rather classy[2].

Steve Suggitt and I were dispatched to a wooden hut of similar vintage, the inventively named Hut 3, where we shared Room 4 - next to the frequently malfunctioning and consistently malodorous loo. We may have not checked our sources thoroughly, but conditions were such that we were both convinced that Alexander Solzhenitsyn wrote many of his best works there. This late and unlamented spot now lies under the Western Campus piazza, doubtless fuelling the growth of the saplings from the soil below.

But four years and two terms in this outpost brought fond memories too. Steve and I used to compete for dissertation students – I cajoled those interested in 'real' human geography (not something extrapolated from a 1930s/1940s German economist) while Steve used more subjective criteria! Bizarrely, some of those first titles live with me – retailing in Blackburn, industrial change in East Lancashire, crime in Skelmersdale and West Lancashire, New Town housing policy. We rotated use of the room, arranging weekly progress meetings around each other's diary, and occasionally even managed to get some work (and a lot of talking) done as co-habitees.

Lectures and seminars invariably took place in the Department – groups of, perhaps, thirty in G3. For me, the timetable consisted of a few first year Introduction to Human Geography lectures – the

---

[2] Compared to the wooden Army huts

'modern' (contemporary economic and social) topics, delivered towards the end of the course – and a couple of Y1 tutorial groups, whose scant knowledge of Physical Geography was nonetheless often ahead of mine. I also shared the second year Urban course (G204) with David Halsall, who picked up the historical, economic theory and transport elements. But some of my favourite teaching in the Department was on G306, Approaches and Concepts. With a few exceptions, typically amongst the mature and/or more able, students mostly loathed the course at the time, though, years later, several of our critics commented much more favourably. Challenging it was, but, overall, it was well taught, and it brought me back into contact with theoretical and empirical work from psychology, sociology and politics.

Most enjoyable of all, however, was a final year option, G310 Social Geography, with sessions on housing, employment, race, class, crime, gender and the like. Typically attracting fifteen or so students a year, most of whom were following related topics for their dissertation, it was possible to combine lecture with discussion group, workshop with seminar, class teaching with a run out in the College minibus. We studied the economic impact on employment change on the Dock Road, post-war housing policy in Everton and Kirkby, segregation in Toxteth and suburbanisation around the Queens Drive. We piled up to Glasgow to look at high-density public sector housing, and nipped to Skelmersdale to consider low density alternatives. Working like this, in the good years, the students acted like mutual discussion and support groups. I still remember the geographers who influenced me as academics and friends, students and staff, and I hope a few of the people I travelled with may have benefited from each other in some similar small way.

But Geography only constituted half of my timetable. Doctoral research on Asian residential and activity segregation meant that first year teaching in Community and Race Relations was unavoidable, and half of the Part II Urban Policy and Race Relations degree depended on contributions from geographers. The work of Andrew Sackville and Nigel Simons, with inputs from Joe McNamara on local policy, stood comparison with the best public policy courses nationally, both in terms of the quality of the sessions and the commitment of the tutors, and the expectations placed on course participants. Given my background, it was a matter of personal regret that I seldom played more than a marginal role in these programmes, though I shared, with David Halsall, Y2 and Y3 UPRR/Geography modules throughout my early years at Edge Hill.

Although appointed as a geographer, I never occupied an office in the Department. When the building was refurbished to accommodate our first ever IT/GIS facility in the early/mid 1980s, Steve Suggitt occupied one of the two new offices, leaving the joys of Hut 3 Room 4 to a sole occupant. And eventually I moved on to the Household Corridor, and progressively to a different set of responsibilities.

Although I consciously published or presented something (and sometimes it felt like anything!) every year from 1974 to 1992, the late 1970s and early-mid 1980s probably represented my peak output as a researcher. Not for the first time, I eschewed the conventional route of doctorate followed by publications. Writing articles was easier; they represented 'bite-sized chunks' of a thesis, could be produced in collaboration with former colleagues and generated far more interest and conference invitations than a doctoral researcher should expect. There were reasons for this – the work we were engaged in was, I believe, interesting, innovative and timely – but I also sensed some curiosity and, even, positive discrimination. Working class lads from the local poly and College of Higher Education, not always playing by the arcane rules of academe, presenting papers at Oxford and Cambridge, working with colleagues from Cornell, being commissioned by the Canadian Government, producing chapters for Academic Press, Allen and Unwin, Oxford University Press, picking up contracts from the Social Science Research Council, having our work published in the Indian sub-continent and covering the back page of *The Sunday Times* Business section, we had a whale of a time.

But the unfinished doctorate began to weigh heavy, and I yearned for a scheme whereby I could submit my collected articles rather than tackle the 140,000-word slog. Such schemes probably didn't exist in the early 1980s and, anyway, almost everything I'd published owed at least as much to the collective endeavours of my supervisors as it did to me. I likened that unfinished thesis – that unstarted thesis - to a cloud on a summer's day, blocking the sun. With the support of colleagues in the Department, we agreed a term of diminished responsibilities (part of a life of diminished responsibilities?) and I wrote in the only way I know. A cheap Amstrad 56mb machine, a 0530 alarm, and a minimum of 2,000 words to screen by 0900, to edit when complete. The middle of 1983 was ruined as a result, but a cloud left the sky late that summer.

I had filleted the PhD – on Asian Residential and Activity Segregation (with case study data from Bradford, West Yorkshire) before I had written it and, after examination (by David Herbert) it became an unopened doorstop. Trevor Jones and I moved on to our next project. In 1989, after numerous missed deadlines, *Social Geography* (Fig 8.1) was published. In many ways the book was written around the G310 Social Geography course I taught at Edge Hill and, doubtless, a parallel course taught in Liverpool. Chapters on Work, Housing, Crime and Disorder, Gender, Race, Neighbourhood and Rural Society sought to take interested geography undergraduates down a path I'd travelled a decade before – a path which introduced them to social and political theory, to the use of historical analysis, to formative case material, but also sought to inculcate a sense of the importance of space and place (as well as class) in the allocative process. Trevor was the driving genius behind the book – though each chapter was written collaboratively, he almost invariably produced a first draft, written on old file paper, A4 envelopes, the backs of menus, leaving me to refine, sequence and justify the more outlandish statements and keep us out of the libel courts. The time I spent on this, the referencing, the indexing, the negotiating with Edward Arnold, probably equated to half of the effort involved, but it was the flair that Trevor brought to the page that made the book a best seller. My only regret was that one of my chapters, on Education, died on the cutting room floor as the 256 pages publishers' maximum was hit.

The book, written in 1987 and 1988 but published in the following year, was perhaps the last item of genuine substance and influence that I produced as an academic. Even today, however, as I find myself on conference platforms talking about widening access, institutional strategy, the statistical bases of league tables and much more, I'm still conscious that I think and analyse as a social geographer. And just occasionally I'd like to step back and write like one too. But, nice though the royalties and the lending rights cheques have been, time hasn't allowed the offer of a second edition to be grasped, and Social Geography drifted out of print last year.

So what took over? The story is a not untypical one. Administration. In the early/mid 1980s responsibility for Edge Hill's Urban Policy and Race Relations degree was thrust into my reluctant hands. Too early and too young, and a seemingly moribund product which the College was contemplating phasing out. I wasn't convinced that I had the academic experience and credibility to cut it, but we did other things well. A great relationship with the students, outreach work in the schools and colleges, open days and conferences, press releases, radio interviews – unusual in the 1980s, if increasingly typical of today. And, of course, plenty of fieldwork, in Newcastle and Glasgow and in cities closer to home. But also, I think, an awareness that a management and product failure at thirty could blight a career for life.

Subject leadership led, almost inexorably, to Head of Department and 'Deputy Dean' status, followed by a spell coordinating Policy, Planning and Development across Edge Hill, around the time of incorporation. The 'Peter Principle' really kicked in 1992, when I acquired the role of Director of Resources, initially covering for an unwell appointee, and twelve months later I found myself acting as Director and Chief Executive – too young and too raw again, but with a strong and supportive team to guide me in that first year or two.

Geography teaching inevitably drifted south in the late 1980s and early 1990s, as a half timetable became five slots on G306 (Approaches and Concepts) and my Social Geography option (G310). I still wanted to get geographers thinking outside of the disciplinary box, reading psychology, sociology, political theory, and bringing their awareness of space and place to the analytical frame. My Social Geography wanted to understand complex social, cultural and economic phenomena – race, class, work, housing, crime, education, gender – and it wanted to recognise that key attributes and characteristics of society could not be wholly understood from one perspective alone. But I was equally passionate that the perspective of the geographer offered insights that stood comparison with those emanating from any other subject. True, we didn't have the range of theoretical constructs developed in the political and social science, but we often had a better sense of how such theories translated on the ground.

At the time, I was often asked whether my social geography was a Marxist geography. The answer was always 'no'. No one political theorist, least of all one writing more than a century earlier, could explain all aspects of social structure, just as no single discipline could. But I do think my social geography was, initially at least, a structuralist social geography – one that sought to understand power and influence and how this was played out 'on the ground'. This interest in the local manifestation of power added an increasingly strong managerial perspective to my interests and my research. Often belittled, context-rich managerial geography, when built on secure theoretical understanding and an awareness of the limitations on individual actions and influence, was also an excellent teaching tool. The local and the familiar could quickly and readily be used to build broader understanding. And it influenced my own research output too. In Bradford, for example, I sought to distinguish the respective influences of social, political and economic position and the actions of local estate agents on black and Asian segregation – and simply proved to myself the complexity of real-world decision-taking relative to the determinism inherent in much modelling.

But then social geography left me behind. Comfortable with much of the work of urban and social geographers published in the 1970s and 1980s, I did not share the paradigm of the early 1990s - a cultural geography, a phenomenological geography, a geography which sought to focus on the individual, in my view without due cognisance being given to the constraints within which we all operate. I also felt that much of this geography came from the political right and, whilst I did not wish to deny that we all have some capacity for self-determination, nor did I wish to underplay the extent to which, knowingly or unknowingly, explicitly or implicitly, our actions, our life chances, are determined by our socio-economic status, our gender, our race, our power (or the lack thereof). And a wander around the Geography shelves of an academic library, thumbing the books of the last decade, do little to convince me that my part of the discipline has the drive and the influence it had in the past.

Epilogue

So, whither geography? From my first days in the sixth form to today in front of the keyboard, I've been convinced that geography is the best undergraduate discipline. Multidisciplinary before the term was invented, a geographical perspective can inform any topic, from D H Lawrence's Nottinghamshire through to an Arctic glacier. It requires the ability to write and the ability to work with data. It combines the practical with the theoretical, the outdoor with the indoor, the human and the physical. It demands an understanding of others' disciplines too – from physics and chemistry on one side of the spectrum through to sociology and psychology on the other.

But there are enormous challenges ahead. Vocational before its time, geography wouldn't sit on the top of anyone's list of vocational degrees today. And yet this is primarily a presentational issue – the learning and life skills a geographer acquires stand comparison with those of any other discipline. And geography is in danger of falling out of the school curriculum. Wrapped up in the dreaded 'humanities' label in Primary schools, barely an option post-fourteen in many Secondary establishments, challenged by the new fourteen-nineteen vocational curriculum, its future is hardly secure.

And too few students come to Geography for the first time at sixteen. But, if Business or Psychology or Sports Science can be studied *ab initio* at AS level, why not Geography too? And is a Geography or related A-level an absolute necessity for University study? I don't think so.

Yet, whilst I would argue that geography is the ultimate integrating and synthesising undergraduate discipline, I'm not convinced of its status as a postgraduate subject in its own right. By this stage the space/place justification seems weak against the established body of theory that characterises most disciplines. Within a day of starting a doctorate I was aware that I would read and depend more on theory and evidence developed in the other social sciences than I would on my own. And yes, I always felt that, as a geographer, I had something to bring to the debate, a perspective, an approach, but rarely a full answer.

But most of all, we have an image problem and a marketing problem. Insufficient people know what geographers can do, what skills geographers have, and the lazy comedy stereotype of cord jackets and leather patches (I've never owned either!) is the most readily recalled image in many people's mind. With 2006, tuition fees and a highly competitive market for undergraduates just around the corner, the discipline has work to do – and this applies equally in its local manifestation at Edge Hill. After two and a half decades associated with a high quality, committed and hard working team of colleagues and friends, I hope that the future is a healthy one, and would want to play my part in securing that.

A Bright Window

Gregg Paget

Edge Hill University College made a great impression on me at a time of significant personal and professional change. I was for a far too short time of the 1981 to 1982 academic year on an attractive university campus on the edge of a Lancashire market town next to the countryside having just come from the Land Economy Department at Aberdeen University where I was for seven years and as it turned out on my way to the Department of Environmental and Geographical Sciences at Manchester Metropolitan University where I remain after twenty-two years.

My perceptions of Edge Hill College, looking to the past, are formed by real achievements and pleasant nostalgia. The real achievement, one of my most important, was the exciting and most challenging task of taking up my first full-time lecturing post which provided a positive foundation for my subsequent lecturing career. It was a 'shock', after my Research Fellow post at Aberdeen University with the luxury of well structured research work patterns with occasional lectures, to face the speed, complexity, tight deadlines and 'thinking on your feet' parts of full-time lecturing. The introduction to closer and more intense involvement with students was a new experience which was most successful. Lecturing at the time was a difficult yet successful and satisfying challenge that I personally and professionally gained so much from.

The working atmosphere in the Department was excellent with friendly and most helpful fellow lecturers – John Cater, Julia McMorrow, Steve Suggitt, Vic Keyte, David Halsall and Head of Department, Derek Mottershead. Nigel Simons, the lecturer that I was replacing while he was on an academic sabbatical (where have those gone?) had regular contact with me and he and his family looked after me very well. And there was this amazing woman, Joan who turned her enthusiastic hand to just about everything – departmental administrator, secretary, technician and morale booster.

I enjoyed teaching the classes in Geography and Social Policy with particularly good responses from the students. Some nostalgia here or maybe I am a field trip 'junkie' but I did complete two enjoyable and most interesting residential field trips- one to the Gower Area of Outstanding Natural Beauty (AONB) in South Wales and the other to the Isle of Purbeck AONB in Dorset.

The students were brilliant - cooperative and friendly. I thought the students performed particularly well – participating in class, interested in their work and producing good standard coursework and examinations. I was particularly pleased by the performance of the weaker students - the staff got a lot out of them. Looking back from my present perspective, I think this good output was particularly enhanced by the close community atmosphere (staff and students) of Edge Hill and the assessment quality monitoring by Lancaster University. As a nice bonus the mature students, some of whom I am still in contact with, were most helpful in contributing to a good social life at College, in Southport and in Liverpool.

I must say something about living in Southport as it made a particular impression on me. At first I found it odd that the students, not in halls of residence, lived and socialised in Southport and like myself travelled into Edge Hill each day by bus. Of course the attraction was the cheap and available housing (that is why I was living in a bedsit in a most notorious house). There were also the seaside resort attractions even in the winter. I remember the winter well - one of the coldest on record - with snow and ice and a large proportion of the Marine Lake frozen over. When the mild weather returned I enjoyed cycling through the surrounding Lancashire countryside. I loved Southport and the easy access to Liverpool which I enjoyed very much - the plays, concerts, pubs, restaurants and the tremendous architecture. Liverpool was certainly a great place for pursuing my academic interest in urban planning. Today, I regularly take Manchester Metropolitan University students on field trips to Liverpool for urban planning, heritage conservation and urban regeneration subjects. Also the Edge Hill experience, and especially Nigel Simons, introduced me to some of my best existing coastal zone management field visits, namely the Southport beach, the Sefton Coast sand dunes and the Ribble Estuary.

Edge Hill University College made a significant impression on me and by doing so laid some important personal foundations. These influences include the successful start of a university lecturing career, the academic and personal value of a small close working department, the unique academic community value, for both students and staff, of a modest sized self contained College and most importantly the good work that the Edge Hill community got out of its students. As I achieved so much and had so many positive experiences in such a short time at Edge Hill College, it surely was a bright window.

Geography at Woodlands

Mike Pearson

Edge Hill's Woodlands Campus

The Woodlands Campus in Chorley has been an important outpost for Edge Hill since the early 1980s. The site was used during the Second World War by the Royal Ordnance factory at Euxton as hutted accommodation for female munitions workers, and the concrete bases of some of these 1940 huts are still visible under the bushes surrounding the car parks. Lancashire County Council owns the site and buildings at Woodlands, which were built in 1975 as phase one of a new Chorley College of Education. Events rapidly changed the plans and Chorley College along with Poulton College of Education was merged with the new Preston Polytechnic in the late 1970s as part of the cut back of Teacher Training Colleges masterminded by the then Secretary of Education, Shirley Williams. Preston Polytechnic used the Woodlands Campus for in-service teacher education and for part-time BEd degrees, which proved very popular with teachers, as many had been trained without taking a full degree course. When, in the early 1980s, Preston Polytechnic decided that teacher training and education was not one of its priorities the

Woodlands Campus was offered to Edge Hill, which initially took a cautious line and negotiated priority use of some rooms for in-service work with teachers. This left the cost of running the centre with Lancashire County Council.

I had joined Chorley College of Education in 1965 after teaching geography at Dartford Grammar School for eight years. At the merger of Chorley and Poulton Colleges with Preston Polytechnic, a strong Geography Division was set up, under the leadership of Howard Phillips who had been Head of Geography at Poulton. In the 1978-1979 academic years, I was given a sabbatical year to take the MSc in Urban Studies at Salford University. As I was interested in in-service work with teachers, I took the initiative to get involved with in-service courses at Woodlands, alongside my academic lecturing at Preston Polytechnic.

Curriculum change in the air

Exciting things were happening in the in-service field at that time. Farsighted Lancashire County Advisers were engineering a complete overhaul of primary education. One of the changes was to replace rather woolly 'topic work' with a more structured history and geography curriculum. A Senior Adviser, Ralph McMullen, organised two Lancashire primary head teachers' conferences at the Sheraton Hotel in Blackpool in 1979 and 1980. One working party for history and geography consisted of thirteen heads, one adviser and two specialist consultants – Geoff Timmins for history and myself for geography. Together we produced working papers for consultation, not without some strong opposition from some heads who liked the idea of mixing geography and history and other subjects into a broad environmental education package. In the event, Lancashire was anticipating the dramatic changes brought about by the introduction of the National Curriculum ten years later. Because Geoff and I were generating a lot of in-service history and geography courses at Woodlands and at other centres like Blackburn Curriculum and Professional Centre we were approached by the Head of In-Service Studies at Edge Hill to join the Edge Hill staff, which I did in 1982. For two or three years I persuaded Lancashire County Council, Edge Hill and the Polytechnic to agree to a quid pro quo situation whereby Geoff Timmins (Preston Polytechnic) contributed to some Edge Hill courses at Woodlands and I in exchange lectured on the urban geography course at the Polytechnic.

Part time in-service degrees

The part time in-service BEd degree started in the mid 70s with Chorley College and was taken over by Edge Hill in the early 80s, until superseded by the part time BA in the early 1990s. The degree was fifty per cent educational theory and practice and fifty per cent main subject work, where one of the most popular options initiated by Bob Wilson, a geographer at Chorley College, was entitled *Urban Environmental Studies*. This was taught by an art tutor (Tom Titherington, Head of Art at Edge Hill College) an historian (Geoff Timmins) and geographers (Bob Wilson and myself from 1982). Urban environmental studies was an exciting cross discipline course to teach and we think that the teachers who took the degree enjoyed and benefited from the specialist knowledge and the enthusiasm for their subjects of the tutors. There was a lot of interchange of ideas between us especially on fieldwork excursions. I remember Tom Titherington opening all our eyes by getting us to really look at the splendid early nineteenth century architecture in central Liverpool, and visiting the stunning cast iron interior of St George's Church in Everton and the amazing architecture and lay out of Port Sunlight. Since then, I have attempted whenever possible to take my student groups on a field trip to Port Sunlight!

The Environmental Education Diploma

The Diploma in Advanced Studies in Education (DASE): Environmental Education was another very successful advanced course for teachers, which Poulton College of Education had initiated. The course was transferred to Woodlands on the demise of Poulton College and it was run by the same tutors as the Urban Environmental Studies course – Tom Titherington, Geoff Timmins and myself. It was available as a full time one-year course or a part time two-year course. As teachers found it increasingly difficult during the 1980s to get their local authority to agree to a one-year secondment to do the full time course, groups tended to be quite small. My records show six teachers completing DASE on a full-time basis in the 1985 to 1986 academic year.

Short courses and conferences

Woodlands was (and is) an excellent centre for short courses and one-day conferences. Many teachers who had completed interesting surveys or fieldwork with pupils or experimental work in the classroom passed on their enthusiasm and expertise to scores of other teachers at one day geography conferences. The idea of teachers talking and discussing with teachers about their work and field studies made for lively sessions in the conference halls in Woodlands. Sandra Skinner (Senior Lecturer at Edge Hill) told me recently that as a primary teacher she used to look forward to the fun and interesting ideas generated at those primary geography one-day conferences. Some of these events attracted big numbers: 140 primary teachers attended a conference in April 1991.

Over the years I have built up a range of contacts of inspiring speakers and seminar leaders for conferences and short courses. Among those who made an impact on teachers' thinking in the 1980s and 1990s are the following. Jackie Micklethwaite, the Education Liaison Officer for the Youth Hostels Association, encouraged many teachers to try out field study work based from youth hostels. Julie Smith, Education Officer for the Mersey Basin Trust, raised teachers' awareness about issues of conservation and our changing waterways. Hilary Glendinning, Head of St Michaels C.E. Primary School in Kirkham, talked about geographical expeditions with her senior pupils to France, the Netherlands and Germany and enthused about the European dimension to the primary curriculum. John Logan, Education Officer for Christian Aid in the North West, was especially good with small groups, having been a primary teacher himself. He used artefacts he had brought back from Sri Lanka to alert teachers to the cultures and life of people in developing countries. Education Officers from agencies like the Global Education Centre in Preston, The Tidy Britain Group (Wigan Pier headquarters) the Lake District National Park (Brockhole centre) and the RSPB made strong contributions to various geography and environmental courses. Two primary teachers who came on my history and geography primary course in the early 1980s made a considerable impact later in the field of geographical education. One was Angela Milner, who went on to lecture in primary geography at St Martins College and is now Dean of Primary Education at Edge Hill College. Her articles and books on early years geography written for the Geographical Association are highly regarded. Joanna Hughes completed the DASE: Environmental Education in 1987 and went on to become a lecturer in primary geography at Liverpool University Education Department and was co-author of the Ginn Key Stage Two Geography books.

Updating geography courses

Courses in updating geography for secondary teachers were frequently held at Woodlands in the 1980s and 1990s, and I received a lot of support in running these from the Geography Adviser for Lancashire, Bill Graham in the early 1980s and later Barry Piggott. Other humanities advisers, especially Bruce Hardman (Bury) Ray Bradburn (Bolton) and Neville Clarke (Wigan) were especially helpful. Some of these were 'twilight courses' running from 3.30pm or as soon as teachers could get out of school to 5.30pm. For example, one entitled 'Geography in the Lower Secondary School' ran for eight sessions in 1984. Other more lengthy courses such as the DES/Regional courses ran for five full days spread out over one term. One of these in the autumn of 1984 included inputs from two HMIs, David Lewis, Geography HMI based in Manchester and Trevor Bennetts, the chief Geography HMI. Edge Hill tutors who lectured on this and other courses included Dr John Cater, Dr Derek Mottershead, Nigel Simons, David Halsall, Paul Gamble while I was the coordinator. Outside speakers at various conferences included Mr R A German from the Commission for Racial Equality, Ashley Kent, University of London Institute of Education, Rex Beddis, author and Humanities Adviser, Avon, Rex Walford, Cambridge University and David Waugh, author. It is not surprising that Woodlands became quite well known at that time as a key centre for geographical in-service work!

Writing textbooks

Inevitably the geographical web you create has the habit of drawing you into new directions! Thus when Derek Mottershead passed on a letter from Arnold Wheaton, the Leeds publisher, asking for someone to lead a team of primary teachers to write a new series of primary geography books, I leapt at the opportunity. The result was the publication of the *Into Geography* series, which comprised four pupil books, and two teacher workbooks aimed at seven to eleven year old pupils and their teachers. The co-authors were Steve and Patricia Harrison who at the time were primary teachers. Steve went on to become a Senior Adviser at Folens publishers. Patricia became an Adviser for Knowsley and a lecturer at Liverpool Hope College and is currently a consultant for the TTA and DFES. The second edition of *Into Geography* for the National Curriculum gained the silver award in 1988 from the Geographical Association with the citation that it 'has been judged to have made or likely to make a significant contribution to geography.'

In 1990 an editor from Ginn Publishers wrote asking if I could bring together another primary team to produce Ginn's Key Stage Two Geography books to meet the demands of the National Curriculum. Three of the Ginn authors were Edge Hill tutors at

Woodlands, Allan Smith, Rosemary Rodgers and myself. The fourth author was Joanna Hughes who was lecturing in primary education at Liverpool University. The Ginn Books also received the Silver Award of the Geographical Association in 1993.

I was pulled into another time-consuming task when I invited Malcolm Renwick to one of the secondary geography teachers' conferences at Woodlands. Malcolm was the chief examiner of the Geography for the Young School Leaver (GYSL) Schools Council Project and GCSE syllabus. Malcolm quickly roped me into the GYSL team to act as Course Work Coordinator for the GYSL schools in the North West. As more schools joined GYSL (now called the Avery Hill syllabus) I eventually was acting, by 2002, as a consultative moderator for some fifty secondary schools across the region. Meetings of the Avery Hill consultative moderators took place three times a year in a redundant primary school, Gerridge Street, in London which ILEA had made into a geography and environmental studies teaching centre. The director of the centre was John Westway, who was the Avery Hill Consultative moderator for schools in the London area. John is now the chief officer for geography at the QCA.

The Geographical Association Diploma

A very special course for Bolton Metro teachers came into being during a conversation between Roy Bradburn (Humanities Adviser for Bolton) and myself at Woodlands. The idea was to run a Diploma course in geography for primary and secondary teachers to be validated by the Geographical Association (GA) and supported by backing from Edge Hill and the Bolton Metro education authority. As far as I know, this was the first time the GA as a professional body had agreed to a professional Diploma course in its name and this was largely due to the enthusiastic support of Eleanor Rawling who was GA President at the time. The course was run as a two-year part time course from 1988 to 1990. The External Examiner was James Price, Head of Geography at St Martins College. Ten teachers (five primary and five secondary) completed the course and Eleanor Rawling presented their GA Diplomas on 25 April 1991 at the Bolton Teachers Centre.

My work for the GA goes back to the early 1960s when Molly Long (then President of the GA) invited teachers who were keen to support the GA to pin up senior pupils' fieldwork exhibitions at the annual GA conferences then held at the London School of Economics. In 1967 I became Hon. Sec. of the Ribblesdale Branch and organised all the meetings, which mainly met in Blackburn

Teachers Centre, until 2002 when Dr Charles Rawding took over the running. By the 1980s most meetings took place at the University in Preston, though the Regional Final of the GA's World Wise Quiz for some years took place at Woodlands in the 1980s.

Probably the most contentious geography conferences at Woodlands took place as the National Curriculum was being created. Eleanor Rawling, who was on the committee writing the draft and final versions of the National Curriculum for geography, used the GA network to get responses from geography teachers across the country. Meetings were convened by the GA in each region. Woodlands was the venue for the North West Region, and I remember a very stormy meeting with a lot of sharp questions asked from the teachers' discussions held during the day. Another consultation meeting was held at Woodlands to discuss the final draft of the National Curriculum which had taken into account the objections and teacher comments from the earlier meetings across the country.

First retirement

I 'retired' from full time lecturing in 1992, but as a consultant in geographical education continued to work hard to support the geography network across the North West. I lectured in primary geography for several terms at St Martins College, Lancaster and also for several semesters at Liverpool Hope University College. However my main lecturing work over the 1992 to 2004 period was with Edge Hill.

The Woodlands part time degree evolved in the 1990s into the part time BA mainly for nursery nurses who were keen to become primary teachers. I helped Allan Smith who was the BA Course Leader to write the two geography units of the degree; one urban geography and the other rural geography and the overreaching environmental studies units, which formed a central role in the degree. I also lectured for several years on the degree course especially the geography and environmental studies units. Several geography tutors from the Ormskirk Campus helped me by contributing lectures for the geography units, especially Vic Kyte, David Halsall and Sylvia Woodhead. In return I did some urban geography lecturing for the geography department at Ormskirk (Fig 8.2). This has continued up until the present with inputs for Taz Shakur's Cities in Transition course. I have also helped out at times with primary geography courses on the Ormskirk campus for the students training to be teachers.

Fig 8.2 Chinese planners making friends with Mike's dog. Chinese people do not in general like or trust dogs, which in China are either very fierce or eaten. They were at first very frightened of Mike's docile Labrador, then became proud of their increased confidence.

Source: Elizabeth Pearson

20-day courses for primary teachers

Another highpoint in the 1990s was the introduction of 20-day courses to help primary teachers improve their knowledge of foundation subjects like geography. As a part time Edge Hill tutor I was able to organise and lecture on these courses for four metro authorities during the 1990s - Bolton, Bury, Rochdale and Wigan. Being a fieldwork geographer at heart, I included a residential element in each one. For Wigan teachers we used the excellent Wigan owned field study centre overlooking Lake Coniston in the Lake District with help from the Centre Warden and Neville Clarke the Wigan geography adviser. For Bolton, Bury and Rochdale primary teachers we used Grasmere in the Lake District as our centre, as it provides all the necessary themes and locations for a settlement study, farm visits and safe river studies.

Fig 8.3 Jeremy Krause, GA President presents Honorary Membership of the Geographical Assocation to Mike Pearson, at UMIST in 2002

Photo Ray Liddy

GA Regional coordinator

While Eleanor Rawling was GA President she had the bright idea of creating a network of regional coordinators for the GA. I was roped in for the North West. The scheme was set up in 1993 with two pilot regions, mine in the North West and Mike Hillary in the Southern Region. This developed gradually into a team of six coordinators during the 1990s. I sent out newsletters of all in-service activities and events across the North West Region to about a hundred key people involved in geographical education. I also supported one-day conferences across the region by selling GA books to teachers, advisers and others. I 'retired' for the second time in 2002 and handed over my role as NW Coordinator to the staff at Manchester Metropolitan University. That year the GA Annual Conference was held in Manchester and I was very touched to be made an Honorary Member of the Geographical Association by the President, Jeremy Krause (Fig 8.3).

Inevitably in writing what I can remember of the geographical work at Woodlands over the last twenty-five years there are many courses and events which have been missed out. I may well have omitted to mention some of the many helpful colleagues who have made contributions on the many courses over the years. Apologies if inadvertently anyone has been left out.

Real World Geography

Nigel Simons

This contribution attempts to explore the relationship between academic geography and the 'real world' over a period of almost three decades. In the process it refers to continuing debates about the application of geographical knowledge and skills and illustrates some of the ways in which geographers have sought to demonstrate the relevance of their discipline through teaching, research and knowledge transfer activities. Much of the material is inevitably drawn from personal experience, and reference is made to the development of geography both at Edge Hill and more widely.

The relevance of Geography and Geographers

In truth, geography has always made a contribution to policy and practice, although the 'visibility' of this contribution has varied. For a discipline devoted to the understanding of both the natural and human worlds, and to the analysis of the ways in which they interact, this is perhaps no great surprise. Yet for many students of the subject and for the wider public, this apparently obvious fact has not struck home. Few professional geographers will have escaped the uncomfortable experience of being told by an acquaintance that 'geography was my least favourite subject at school', or that it was 'boring' or taught by everyone's traditional image of an old fashioned schoolteacher. Clearly, for the future health of the discipline, this is not particularly encouraging.

Perhaps the problem lies in our collective inability to recognise the contribution that has been made, or in our reluctance to celebrate geography's many successful engagements with policy problems.

Explicit recognition of geography's potential to influence the policy process did not really materialise until the 1970s. Partially as a reaction to the self-serving introspection of the 'spatial science' era, some geographers began to advocate a more relevant role and an active engagement with emerging policy problems (Smith, 1973; Coppock, 1974). Thought was also translated into action, and new policy-related research and teaching began to emerge in fields such as resource management and social geography. The development of geography at Edge Hill through the 1970s and 1980s in many ways mirrored these trends, and paved the way for a variety of initiatives in course development, research and knowledge transfer in the 1990s.

From the perspective of the early twenty-first century, now is a good time to take stock of these developments and to consider how they have positioned geography and geographers. We need to ask ourselves how well geography has applied its core knowledge and skills to the development of policy and, perhaps just as importantly, to its implementation and evaluation. Equally, it seems appropriate to assess geography's current position in an increasingly market-led higher education system, and to draw some tentative conclusions as to its future prospects.

Geography and Resource Management

Geography is a notoriously diverse discipline, characterised in the post-war period by increasingly disparate groups of specialists focused around the natural sciences (for example, geomorphology, climatology, biogeography) and the social sciences (in areas such as historical, social and cultural geography). This diversity has often been considered divisive and damaging, particularly in the context of the apparent rift between physical and human geographers. Yet there has always been a third strand in geography around the human-environment interface which allowed the two main traditions to work together and to understand the complex interactions between environmental conditions and human activity. Here, perhaps more than in any other branch of the discipline, lies a real strength and potential to influence future policy and practice. Even in the 1920s, attempts were being made to recast geography as the science of 'human ecology' (Barrows, 1923) and individual geographers were increasingly influential in advising government agencies (especially in the USA) on key problems such as river flooding.

Whilst the efforts of individual geographers kept this approach alive and relevant to policy, it was not until the 'environmental agenda' began to take off in the 1970s that it experienced its full realisation. O'Riordan (1971) offered an important early review of this (mostly US-based) work in resource management, and geographers developed particular strengths in, for example, the analysis of natural hazards such as flooding, earthquakes and coastal erosion. Simultaneously, there was a growth of interest in the management of key natural resources such as water and minerals, and in industrial problems including pollution of the air, land and water. All of these growing environmental problems appeared to match well with the knowledge and skills of the geographical community – an understanding of environmental systems, an ability to monitor and measure environmental trends, and an appreciation of human behaviour and responses. With the recognition of this 'fit', geographers produced an impressive range of applied, policy-relevant studies combining scientific analysis with insights into institutions and practitioners (for example, Parker and Penning-Rowsell, 1980 on the UK water industry).

Geography at Edge Hill was not slow at responding to this trend, offering its students early insights into both the theory and practice of resource management. Combining the interests of both physical and human geographers, a popular second year course entitled Resource Management ran for several years from the late 1970s[3], introducing core concepts and applying these to a wide range of current issues such as land use change, nature conservation, energy and mineral resources and the management of specific environments such as the uplands and the coast. A growing literature of both academic geographical texts and policy documents (such as planning reports, working papers and management plans) kept students engaged with the contributions of geographers and with the work of practitioners in the 'real world', and there was ample opportunity to use video material and to link class-based teaching with fieldwork. Such a course was inevitably a broad survey of many topics, rather than a detailed examination. Having interested so many students, it was necessary to go further by offering specialist options in the final year. These expressed the particular research interests of staff at the time, and included a course in Coastal Zone Management which expressed my own active work at the time (Simons, 1982). All of these options expressed the strength of combining physical and human geography in the analysis of issues which remain highly topical today: coastal flooding, water supply, and drought management being particular examples. As we shall see later, these developments subsequently bore fruit in the development of entire degree programmes in the early 1990s.

There was perhaps one weakness in this work, in that students were rarely fully formed practitioners at the end of their course. They had a strong academic background in resource management, but had

---

[3] The Environmental Resource Management course continues today, and is still popular with students (SW)

not always acquired the necessary technical skills or encountered practitioners at first hand. Some corrected this deficiency through their dissertation work, whilst others went on to study at postgraduate level in specialist areas such as environmental assessment. A healthy number eventually entered environmental practice in local authorities, government departments or industry, and some became consultants – demonstrating that employability and professional relevance are themes with a longer history than many might believe.

Geography & Urban Policy

The search for relevance and a connection with the world of public policy was not restricted to the environmental arena. From the late 1970s, social geographers were strongly committed to understanding the allocation of scarce resources (such as housing and health care) within the contemporary city, and to revealing the disadvantaged position of certain areas (such as the inner city) and social groups (including minority ethnic groups). Evidence of urban and regional inequalities was compiled using traditional geographical skills in data acquisition and analysis, and the discovery of these inequities led some to claim the birth of a new field of 'welfare geography' (Smith, 1977). Merely describing the existence of inequalities was not enough, however. There was a real hunger for explanation, and for an opportunity to diagnose the impacts of government policy. The explanation of inequality drew social geographers into an extended engagement with the broader social sciences, particularly the 'grand theories' of Weber and Marx. Increasingly, there was a recognition that geographical distributions of poverty, ill health and bad housing could only be understood against a background of social and political structures. This growing preoccupation with broad structural factors also increased interest in the ways in which public policy towards key areas and social groups was made and implemented. Against a background of the Thatcher government's attempts to 'roll back' the frontiers of the public sector and introduce the disciplines of the market, social geography became increasingly critical, politically aligned and determined to seek ways to remedy what were seen as injustices.

Against this background of academic and political change, human geography at Edge Hill found an outlet in cross-disciplinary collaboration. Working strongly with a group of like-minded social scientists, we created what was to become the BA Urban Policy & Race Relations programme. This brought together an assortment of staff with surprisingly common interests and reference points, and began a stimulating 'voyage of discovery' into the uncharted waters of multidisciplinary course design and delivery. The human geographers (including myself and John Cater) worked alongside several social scientists to create a common core of modules outlining British urban development, the urban policy process, and policy as it affected local communities. Delivery of these courses was by genuine team teaching, and this created an opportunity for all involved to learn from each other whilst teaching the students. The degree attracted a fascinating assortment of students including significant numbers of mature entrants and some with previous policy experience, and the dynamics of class teaching were both unpredictable and stimulating. A characteristic of the course was the extraordinary commitment and passion of the students, and many went on to apply their learning actively in local authorities, voluntary bodies or in community settings.

This experience was extremely positive for a number of reasons. Firstly, it was for all concerned a taste of the potential of multidisciplinary collaboration, and as geographers, we absorbed many valuable insights from our colleagues and the students. Secondly, we were working on a course whose entire focus was on policy and practice as it affected different groups and locations. This focus was reinforced by regular contributions from practitioners and by field visits (very much in the geographical tradition) to major cities such as Bradford and Newcastle. This increased our interest in practice 'on the ground' in an influential way, and the vocational path taken by many graduates was a vindication of our approach. Finally, it paved the way for later policy-focused work and was a key forerunner of many initiatives in course development, research and knowledge transfer.

Geography in a Multidisciplinary Environment

Our work at Edge Hill in the urban policy field taught me a number of important lessons about relevance, skills development and the need to engage with practitioners. As I moved to a new post at the University of Central Lancashire in the early 1990s, I doubt that I realised just how much had been absorbed or how I would deploy that learning.

The University (which came into existence in 1992 after I arrived) was a much larger and more diverse institution. I had been recruited as a geographer, but did not realise until later that my new employer had other ambitions. The agenda, as I quickly discovered, was to bring together the University's environmental expertise from many different disciplines and apply this to course development, research and income generation for a growing market. There followed an intoxicating few years of new course development and external work that brought rapid growth and success, to the point where we had created an entirely new University department in the space of around eighteen months. All of this was built by a small group of highly committed geographers, operating as the core of a diverse multidisciplinary group embracing ecologists, lawyers, chemists, engineers and biologists. Here, geography (with its traditional breadth across the natural and social sciences) could operate to cement the various specialist contributions into a coherent whole.

The focus of all this activity was the development of a new BSc programme in Environmental Management – one of the first in the country at the time, soon to be followed by many more. Underpinning this key development was a clear philosophy, building on much of my previous experience at Edge Hill. The main components of this were:

- Geographers had a major role to play in launching and supporting environmentally-focused work
- Such activity was necessarily multidisciplinary, and spanned the natural and social sciences
- Students should be exposed to the 'real world' of policy and practice as much as possible, and should leave the University with relevant vocational skills
- Taught courses in environmental subjects had to be underpinned by active applied research and technology transfer, opening up opportunities for students to engage with 'live' problems
- International collaboration was necessary to open up global perspectives and comparisons, and give students overseas fieldwork opportunities
- Teaching and research in sustainability should be underpinned by good institutional practice – the University needed to ensure that its own treatment of the environment was aligned with best practice

Much, if not all, of this approach was secured during the 1990s expansion of the Department. The geographers proved their worth as leaders of multi disciplinary initiatives, although there were often some interesting moments of incomprehension between staff who spoke and thought in very different ways. Student numbers grew at a headlong rate for several years, and graduates left with skills that placed them well for future employment – the majority entered the environmental sector and many progressed quickly to more senior posts. We built an impressive network of professional contacts, encouraged practitioners into the lecture theatre, and undertook a number of innovative projects with local industry (Simons, 1992). Research grew quickly, and there was a strong policy dimension to this – I became actively involved in work into pollution modelling and transport policy, the impacts of environmental reviews in various organisations and, most recently, the implementation of local air pollution control (Simons & Slinger, 2004). International work also flourished, with funded collaborations in Hungary and India, course development in China, and research and fieldwork in Kenya. We also broke new ground by undertaking the first environmental review of any UK university (Merritt, 1993) and extended this to develop a working Environmental Management System (Merritt, 1994).

Whilst the environmental interest and commitment of the 1990s has waned, much of this work continues to prosper. The underlying philosophy remains robust, and current concerns with skills and employability seem well matched with our original ideals and delivery. Above all, geography and geographers have continued to play a central role in the university's environmental provision. Indeed, as some of the environmental courses have seen their student intakes dwindle, it is geography that has grown to form a stable and lasting base for the department.

Geography in the Twenty-First Century

Much of this contribution could be read as a personal history, making reference to various trends in geography and the wider world of higher education. Hopefully, it offers a little more – particularly some insight into geography's continuing engagement with policy and the challenges that this has posed.

Perhaps the first point to note is that geographers have demonstrated beyond question that they can make meaningful inputs to policy analysis and to practice 'on the ground', and that these contributions cover a wide range from the environmental to the socio-political. Early attempts to make geography 'relevant' were perhaps rather naïve and did not fully connect with practice, but this was remedied in much of the more recent work. In my own experience, geographers began to find their feet in this respect when they became involved with multidisciplinary ventures such as the development of environmental management in the 1990s. In the process, they were confronting issues of vocational skills and employability some time before they became key issues for higher education providers. Curiously, the environmental 'bubble' seems to have burst for now, and students are not always clear about what geography has to offer them in an increasingly competitive, market-driven higher education system. With increasing levels of student debt and intense competition between departments for student numbers, there is a need to identify its 'unique selling points' and play to its traditional strengths, particularly in developing key skills and in offering a broad, flexible curriculum that embraces both the natural and social sciences (Chalkley & Harwood, 1998). As a discipline, geography needs to more self-confident in recognising and articulating its many successful contributions to policy and practice – it will not help us if we know our worth, but the rest of society does not!

References

Barrows, H. H. (1923) 'Geography as human ecology' *Annals of the Association of American Geographers,* 13, 1-14

Chalkley, B. and Harwood, J. (1998) *Transferable skills and work-based learning in Geography,* Cheltenham: Geography Discipline Network.

Coppock, J.T. (1974) 'Geography and public policy: challenges, opportunities and implications' *Transactions of the Institute of British Geographers, 63,* 1-16

Merritt, J.Q. (1993) *Environmental Audit Project: Final Report (2 Vols)* Preston: University of Central Lancashire.

Merritt, J.Q. (1994) *Environmental Management Systems Project: Final Report* Preston: University of Central Lancashire.

O'Riordan, T. (1971) *Perspectives on Resource Management,* London: Pion

Parker, D.J., and Penning-Rowsell, E.C. (1980) *Water Planning in Britain London:* Allen & Unwin.

Simons, N.F. (1982) *Coastal Zone Management in England & Wales, with special reference to East Sussex,* Unpublished PhD thesis, University of London.

Simons, N.F. (1992) 'The environmental challenge,' *Modern Management* , 4 (3) 42-44

Simons, N.F., and Slinger, P.G. (2004) 'The implementation of Local Air Pollution Control in England & Wales: a case study assessment,' Paper submitted to *Environmental Politics.*

Smith, D.M. (1973) 'Alternative "relevant" professional roles,' *Area, 5* (1) 1-4

Smith, D.M. (1977) *Human Geography: a welfare approach,* London: Arnold.

Duplicators and equipment inventories

Joan Swinhoe

I started working in the Geography Department at Edge Hill in 1979. Head of Department was Miss Anna Cooper; the tutors were Andrew Francis, Paul Gamble, David Halsall, Neil Immins, Vic Keyte, Julia McMorrow, Nigel Simons and Steve Suggitt. I was appointed as Technician, but the work I was required to do was mostly clerical. I was there to help tutors, but if Miss Cooper needed me, she had first call. She made sure when I started College that I was introduced to all the other departments. She suggested I go to the Staff Common Room for my coffee so I would get to know other staff. I did this and I made many friends, which I still have today. You bought tea or coffee tickets from the Finance Office, and the Common Room was always full of both academic and administrative staff. Miss Cooper kept very much to herself. She lived in College and was warden of Lady Openshaw Hall (I think). She always lived in College and before that in the school where she had taught. She had a house in Wales, but didn't go there often. After she retired she remained a warden.

I had a manual typewriter, and there was a duplicator where lecture notes were run off. There was one Xerox machine in College and you had to fill in an order form requesting the number of copies and the date they were required by, you had to be quite organised because you usually had to wait three or four days for the Xeroxing to be done. If you wanted notes or portions of maps for fieldwork, they had to be requested days ahead. At the end of the financial year the department received a bill for Xeroxing, and the amount was taken out of our budget. We eventually got our own Xeroxing machine, oh how it made life easier, and of course the quality of notes was so much more professional. There was also a small library in the department that was for tutor use only. Tutors were given a budget each year and they would instruct me what to order. When the books arrived I would catalogue them and put them into the library. When students borrowed equipment I would record it and make sure it was returned in good order, because we had an inventory book where each piece of equipment was recorded and at the end of the financial year the inventory books had to be made available to the auditors. In the early days every item was recorded, tape measures, compasses, ranging poles, but later only the more expensive items were recorded.

I would book coaches and accommodation if necessary for the many field trips. These trips were always in England until Steve Suggitt organised a trip to Norway. The students who went were usually going to do research for their dissertations, and they paid their own expenses. They camped, made all their own food and travelled in a College minibus, which had to be booked months in advance. I used to go with Steve to the supermarket to buy the food. He had the menus worked out and we would fill two trolleys, much to the amusement of the other customers. Very little if any food came back. On their return students would come and tell us of the wonderful time they had had.

Fig 8.4 Edge Hill geographers at Joan Swinhoe's leaving do. From left to right: standing: Peter Stein, Kathryn Coffey, Nigel Richardson, June Ennis, Gerry Lucas, Kath Sambles, Tasleem Shakur, Ann Chapman, Steve Suggitt, John Cater. Seated: Richard Jones, Derek Mottershead, Joan Swinhoe, Paul Rodaway, Sylvia Woodhead. *Source: Joan Swinhoe*

I got to know many students very well, and they would often come to my office for a chat, I think it was easier for first years to ask me questions rather than go to a tutor. Of course once they got to know the tutors it was a different matter. Students often would visit us after they graduated. The department was a very friendly place, and after Miss Cooper retired and Derek Mottershead was appointed Head of Department there was a more relaxed atmosphere. After we had a storeroom, which had been the tutors' 'coffee lounge', converted into a laboratory, tutors either stayed in their offices or came to my office. Many in-depth discussions took place over coffee especially on a Friday morning after the Thursday evening staff cricket match. Nora was appointed as a cartographer, but she didn't stay very long, then Ann Chapman was appointed and she fitted in very well and was a great help to tutors. A technician, Kathy Coffey was appointed, so now our Lab. was operational. We had one Apple computer in the department which was used to varying degrees by all of us. Then we received PCs, oh what fun we had learning to use them. I could now produce beautiful charts of the department's finances. It took me twice as long, but they did look nice. Having student records on the computer was a great help. We had a change of Head of Department, Paul Rodaway, who wasn't in the job long because he was appointed Head of School. However that was after my time because I retired from Edge Hill in November 1995.

John Cater

When John arrived in the Geography department he settled in immediately, in fact it wasn't long before he was treasurer of the tutors' coffee fund, 'snatch, grabbit and run'. He was very popular and a path was beaten to my door by female admin staff and students seeking any information I could give them about him. One geography student caught his attention and after she graduated they married. Students always said how good his lectures were, but he could never stand still while lecturing and used to pace around the room the whole lecture. I was often called upon by students to decipher his comments on their essays; his writing was very neat but almost unreadable until you got used to it. He was always in College very early and most mornings I would come into my office and John would be working, a coffee close by and his books spread all over. He was a member of the staff cricket team and quiz teams, played rugby for Ormskirk and cricket for local team. He was always cheerful, not always complimentary but said without much malice.

Paul Gamble

When Paul was put in charge of Afro-Asian Studies he persuaded Miss Cooper to let me work for him one afternoon per week. I would work in the Afro-Asian library, a small room where the secretary for History also worked. I ordered and catalogued the books, did orders for other items, checked and passed invoices for payment to the Finance Office. Paul took students to Tunisia each year and I would help in the booking of the trip. We always made a display of the photographs which were brought back. At exam time I would collect the students' coursework and make sure it was all correct for the

External Examiners' meeting. I enjoyed working for Paul and he was missed by students and staff when he died.

Vic Keyte

Vic was First Year Tutor. He had a very set pattern to his life. He would cycle into College almost every day, rain, hail or shine. He would arrive between 9.45 and 10 o'clock unless he had a 9am lecture when he would drive in. At 11am he would have coffee and two chocolate digestive biscuits, then lunch in his study at 1pm. He swam two or three times per week; he was a very fit man. He was quite vague about life going on around him, and I usually had to remind him of departmental meetings. Julia McMorrow used to say he was really good on fieldwork, he would explain things to students so well and keep their interest. He wrote a Town Trail on Ormskirk, and produced a very good leaflet on the old houses of Newburgh.

David Halsall

David was Second Year Tutor. He was married to Fiona Montgomery who was a History tutor. He was very keen on steam trains and used to travel all round England to take photographs of them, hence he ran a Transport course. He used to come into College with two or three very full executive briefcases (plastic bags) then would be unable to get into his office because he had left his keys on top of the car. He never threw any piece of paper away; consequently his desk, chairs and windowsill were piled high. He knew where everything was, though sometimes his system didn't work and a search for missing essays took days.

Part III The Future and Beyond

9: Towards the End of the Century

The 1990s marked a new beginning, as well as continuity with the past. It was a time of significant change: a broadening of the range of programmes, the rapid development of a postgraduate research community, and a real vitality of staff and students exploring the frontiers of geography.

Andrew Griffiths offers a southern view of his first 'proper job', recounting the difficult transition from research to teaching undergraduates, whilst Fiona Lewis, continuing the Edge Hill theme of historical geography, remembers with affection the pleasant workspace accorded by geography's separate building. Paul Rodaway recounts the cultural turn in human geography, involving the study of real people and their relationships to the environment, and explaining that post-modern geographies study many aspects of the modern world, from body image to cartoons and videogames. Two research students each explore in different ways the life changing impact and relevance of both human and physical geography research. Nick James reflects on his studies on food security in Zimbabwe: his account shows the complexity of people's relationships with the African environment. Barbara Lang explains how her research on microscopic midges from lake mud is helping to define how the climate changed at the end of the Ice Age.

A Southern View of Edge Hill

Andrew Griffiths

Edge Hill was my first 'proper job' - I was in my mid twenties, and ended my ESRC postgrad grant on farm diversification at Exeter University six months early in order to take up the post. This was lecturing in Human Geography, including courses in Environmental Management and Statistics. The latter was far from my strong point, and I was astonished and somewhat flattered when a student came up at the end of it and thanked me for making stats interesting (only slightly tempered by the widespread acknowledgment that said student was barking mad).

Poking around in dusty archives and talking to Devonshire farmers was deemed good training for lecturing to a room full of eager (and not so eager) students, and although the experience has stood me in good stead, I never felt very comfortable with my lecturing abilities. Despite these reservations some of the students probably learnt something, and at least when I was writing illegibly on the blackboard (it seems like another age.) they could snigger at the sporty ponytail I had at the time. One day I certainly made an impact - striding purposefully into the main lecture room, the door swung back and fell off its hinges with a resounding crash.

Fig 9.1 Andy and the girls: Andrew Griffiths, Edge Hill lecturer 1990-1991 finds he doesn't miss marking essays.

Source: Andy Griffiths

Tutoring students afterwards at Exeter University was tame in comparison: the Scousers and mature students at Edge Hill were great, always questioning and thinking for themselves, unlike the nice Home Counties kids who would sit there and swallow any amount of nonsense. Life under the genial Derek Mottershead was not at all bad, and as a southern boy living in Lancashire the honesty and friendliness of the locals was very refreshing. We bought our first house, in Platt Bridge, for rather less than £20k, and doubled our squad of rescue dogs to a total of four during our stay. Huge piles of essays and exam scripts were a down side of academic life, but the field trips (Northumberland and Glasgow) were fun, and the small team of academics were a great bunch to work with. With Joan Swinhoe keeping a motherly eye on us all, there was a family atmosphere lacking, I suspect, in most academic establishments.

Having failed to finish my PhD, let alone start any other research work, the prospect of years more teaching did not appeal greatly, and so when I was asked if I'd like to return to Exeter University and run the Tourism Research Group I accepted. This led to ten years of consultancy work, supplemented with academic odds and ends, latterly setting up as an independent company (AcumeniA) (Fig 9.1). Among the 150-odd reports authored during this time were the annual 'Cornwall Holiday Surveys', the largest and longest running survey of its type in the country.

However, playing trains has always been my dream, and so when Wessex Trains was created in 2001, based in Exeter, I was able to join as Business Manager for Devon & Cornwall. So now I find myself standing in front of a roomful of people again (Councillors, Users Groups etc) - but thankfully there are no essays to mark afterwards.

Its Own Little Place

Fiona Lewis

I graduated in geography from Manchester University, and had completed my PhD on *The demographic and occupational structure of Liverpool: a study of the parish registers 1660-1750* at Liverpool University in 1993 before I was appointed as lecturer in human geography at Edge Hill. I was appointed at the same time as Paul Rodaway and Kath Sambles, but I lost touch with Kath when she moved to Australia. I replaced Vic Keyte who was retiring, and I took over his course on Heritage Landscapes.

My favourite moments at Edge Hill were the field trips, taking the teaching out of the classroom (Fig 9.2). The students were more motivated then than at any other time, and I remember being amazed at how hard the students worked.

I remember Edge Hill as being a pleasant place to work. The geography department was in its own little space, away from the main building. There was a good atmosphere. Our own place.

I left in 1995 to live elsewhere, but have been back to the area a couple of times, mostly to Liverpool, to work on a book with my supervisor and another researcher. It is an academic history of Liverpool in the eighteenth century. Currently I work as a freelance writer and researcher and a number of my projects have had a historical geography theme. Most of the publications I was working on at Edge Hill have come out since I left; some are coming out this summer. In fact it has taken about nine years to pull together the material for the book about Liverpool.

Fig 9.2 Human Geography fieldwork in The Netherlands. From left to right Tas Shakur, Fiona Lewis, coach driver, David Halsall, Paul Rodaway.

Source: Sylvia Woodhead

I am very interested in a book of this kind as it is very much my subject area. I do a lot of the kind of searches involved, and use amongst other things many of the online sources now available. Tracing people and researching their lives is now far easier and more straightforward than it used to be, although still time consuming. It is interesting to see just how many members of Edge Hill's geographical past have been tempted out of hiding.

The Making of a Post-modern Geographer

Paul Rodaway

Just remember the Geography...

I suppose this is a story about change, and I feel honoured that I was involved in a significant period of change at Edge Hill. I arrived in 1993, having spent five years 'down South' in Bognor Regis at the West Sussex Institute where I had spent much of my time on fieldwork and projects abroad – mainly France, North Africa and especially Romania. There I had learnt all about geography 'at the edge' – taking sixty-five Malaysian (mainly women) to post-Communist Romania in 1990. With tanks still on the streets, all the Malaysian women had to fear was Romanian male students from the Forestry Faculty of the University of Brasov. Later I was again on the edge of a war zone in Slovenia, on fieldwork exploring the countryside and towns near the Croatian border with the guns still firing in 1991. This was not the geography of ranging poles or questionnaires, but the raw first hand experience of 'just being there' and meeting the local people 'making a new geography'.

As you can guess, I am a human geographer. What has excited me about geography is social and cultural change. My two early inspirations were an avid interest in travel and reading the reflections of the American Geographer Yi Fu Tuan (e.g. Tuan 1974). Tuan began as a physical geographer, but early in his career realised his real interest lay in seeking a clearer understanding of our individual and cultural relationship to the environment (see Rodaway 2004). He is one of the few, if only, Geographers to have written a full-length autobiographical reflection on this life and career, *Who am I? An autobiography of emotion, mind and spirit* (Tuan 1999). My career as a geographer has been less illustrious, but has nevertheless taken me on a journey from my first degree at St Andrews University[1] and a doctorate at Durham University, to higher education teaching posts in Liverpool, Bognor Regis & Chichester and Ormskirk. I was a member of the Edge Hill Geography department from 1993-2003.

Today I am the Director for the Centre for Learning & Teaching at the University of Paisley, Scotland, but at heart I am still a Geographer. I am a 'Post-Modern Geographer'. I have an interest in the geographies of our contemporary experience: those 'outside' in our 'themed' and 'commercialised' landscape and those we participate in or 'consume' whilst watching television or going to the cinema, when Internet shopping, in on-line chat rooms or as we learn via a Virtual Learning Environment. I was briefly Head of Geography between 1994-1998, but taught in the department throughout the period 1993-2003. Edge Hill Geography was always an exciting place to be. Here both staff and students explored the frontiers of the discipline and crossed frontiers on numerous field-courses to Europe, Asia, Africa and the Caribbean. If there is one piece of Geography that many colleagues and former students will remember from me, it will probably be my work on cartoon geographies between 2001-2003,

---

[1] Paul Rodaway was at St Andrews 1979 to 1984 when Peter Cundill (See Chapter 7) was there; another of the intricate Edge Hill web of links all over the country

which included a research paper on Apu in *The Simpsons* (2003). I had used cartoons to aid my teaching for many years, but this was perhaps the first time I got a mention in the local press for my research.

On arriving at Edge Hill in 1993, I was thrown in the deep end in teaching social geography, environment & planning, and countryside management. I also soon got involved in teaching the philosophy and methodology of human geography – in many ways my first love. Before I was a Post-Modern Geographer, I was a Humanistic Geographer, and before that some say a Marxist Geographer. My doctorate research at Durham University was thoroughly humanistic - a phenomenological study of the experience of place in a former coal mining village in County Durham. I spent almost eighteen months 'going native' in a participant observation strategy which involved living in the community and conducting what I termed 'group reflection' with a group of local people (Rodaway 1988). We would now call this reflective work 'focus group' research and many Geographers have adopted this approach to getting a greater depth of insight into social and cultural geographies (see AREA 1996). Doing this research was great fun and 'my group' even wrote the middle section of my PhD thesis as a booklet called *Our View of the Valley*. However, this was serious research, pushing back the frontiers of how we do Geography and drew on the diverse ideas as humanistic geographers such as David Seamon (1979) Graham Rowles (1976) and Yi-Fu Tuan (1974). In many ways this work anticipated later Geography by feminist geographers such as Gill Valentine (e.g. 1993) and activist geographers such as Routledge (e.g. 1997). My supervisor at Durham, Douglas Pocock, was also to explore the notion of experiential (and humanistic) research through his work on tourist perceptions (Pocock 1992) and personal experiences of place (Pocock 1996). At Edge Hill, a number of undergraduates explored these ideas in their own research projects into topics such as images of the Lake District, children's geographies, and a study of neighbourhood watch schemes.

In 1993, I had just completed a book entitled *Sensuous Geographies: Body, Sense & Place* (Rodaway 1994). This book (Fig 9.2) started from the premise that Geography begins with the human body, its posture, scale, orientation and mobility, and most importantly its sensory and mental capacities. The book explored the 'sensuous geographies' of sight, touch, hearing and smell. The book argued that the senses should be understood both as sensation (physically) and as meaning (making sense or mentally) and that the senses 'structured' our understanding of the world. Furthermore, the book argued that how we use our senses and respond to sensory inputs is both biologically and culturally determined. The book fell into three parts: an exploration of the approaches to the sense and perception in geography, and an introduction to the distinctive character of each sense; a more detailed exploration of the nature of sight, touch, hearing and smell and the geographical worlds they realise for us; and finally a brief exploration of more contemporary 'hyper-realisation' of the senses (post-modern geographies) in themed environments of shopping and leisure, and in television landscapes such as soap operas and dramas.

The underlying inspiration of the book had come not only from my earlier work at Durham, but also more recent research I conducted in the 1990s into the alternative geographies: of the disabled (especially the autistic, the blind and the deaf) of children and the elderly, and of other cultures (particularly the Saami & Inuit of the arctic, Arab and Japanese culture). Particularly formative of my growing interest in this area at this time was reading work by non-geographers that struck me as giving real insight into alternative geographies. Two contrasting examples being John Hull's *Touching a Rock: An Experience of Blindness* (1990) and Edmund Carpenters' *Eskimo Realities*

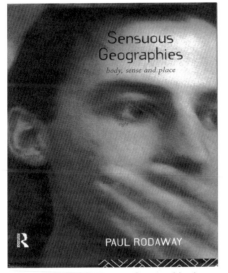

Fig 9.3 Cover of *Sensuous Geographies* by Paul Rodaway

(1973) which each revealed something of a hidden auditory world and the role of hearing in how humans can make sense of a world around them. Many of the topics covered in my book were to be touched on or developed in my lectures in the coming years at Edge Hill. In the first year course, I introduced the notion of body geographies, the geographies of children and the elderly, and the notion of enabling and disabling geographies. In a third year module on *Culture, Technology & Space*, I explored further the ideas about post-modern geographies of retail and leisure environments, and the geographies of film and television. My initial ideas about sensuous geographies were supplemented by ideas from social theory which conceptualised this in terms of 'social spatialisation' and 'the production of space' (Shields 1991, Soja 1989).

When I joined Edge Hill, the department was on the cusp of lots of change. In months of my arrival, three other new tutors, Nigel Richardson, Kath Sambles and Fiona Lewis, had joined the team. Within two years, we transformed the geography curriculum, from one Geography degree to separate programmes for Physical Geography, Human Geography, Environmental Science, Geology, Urban Management and of course Geography. For the first time, we had 'fieldwork modules' – no longer was fieldwork just an add-on. Transferable skills of relevance to employment were made more explicit and many new topics were introduced to the curriculum. With more modules, and more choice, students could now explore areas of geography at the cutting edge of the discipline.

In 1993-1994, my first year, I had a hectic time finding somewhere to live, getting the book completed and off to the publisher, writing lots of new lectures (and marking lots of student work.) and of course there was fieldwork. This was my first visit to the Netherlands. In those days we did not fly the direct route, but instead it was a long coach ride to Dover, a sail across the English Channel, and a long coach ride through France, Belgium and southern Holland to Utrecht. And my first year it rained and rained until the Dutch polders appeared to be floating rafts on the North Sea. Yet, we all fell in love with the Netherlands. Flat and boring? Never. Later years, we let the plane take the strain (to paraphrase the old railway advert) flying from Manchester to Schiphol, and basing ourselves in Amsterdam (and enjoyed some sunshine filled field trips). Dr David Halsall, a long-standing tutor in the department, was an expert on the Netherlands, and especially the Dutch public transport system and directed us around the Netherlands by train, bus and foot-power. Away from the hermetically sealed fish bowl of a luxury coach from England, we began to experience the real Dutch landscape and its people. All the human geographers went to the Netherlands in their second year and would explore a range of topics about the Dutch landscape, people and economy. We explored old and new urban landscapes, industrial heritage, and modern water level control schemes. In the Netherlands, I was particularly interested in the distinctive approach the Dutch have taken to urban design and planning. In Amsterdam, students conducted a study of images of place, exploring the palimpsests of meanings inscribed on and through the urban landscape in the buildings, their architectural styles and changing uses. Most years, we also had the contribution of a local Dutch academic and planning specialist, who provided a key input into one of the evening discussions. A key part of the field-course was student projects and considerable excitement arouse on the final night of the field course when each student group would present their initial findings.

Fig 9.4 Fieldwork in Cuba. Outside the Museum of the Revolution in Havana.

Source: Sylvia Woodhead

The commitment to fieldwork has been a long held tradition in the department. In addition to the formal field courses embedded within individual modules, staff also set up optional field courses to more unusual destinations – such as China, Cuba and Morocco. Perhaps one of the most informative trips was to Fidel Castro's Cuba (Fig 9.3). By this time I was Head of School (1998-2003) and often could not find the time to go on fieldwork, but Dr Tasleem Shakur and Sylvia Woodhead persuaded me to give up those meetings and all that paperwork, and *'get out in the field'*. We arrived in Havana, after dark. It looked so drab, so flaky, so swamped with the wash of the Caribbean. Yet within days, we had grown to love the beauty of Cuba and its friendly people. Having spent much time in Romania many years before, for me it was like a returning to the faded vision of communism.

What did we do in Cuba? In collaboration with colleagues from the Technical University, Havana, we explored revolutionary Cuba both through the eyes of Cubans and our own personal observations of its achievements and challenges. The field course included: visits to the Museum of the Revolution, to a city hospital (which had originally been built as a bank.) an environmental project outside of Havana on a former Cotton Plantation, an exploration of Old Havana and the impact of 'Americanisation', a visit to the University, and for some of us a visit to a Cuban cinema to see a comedy about Cubans exploiting tourists (in Spanish of course). Students also spent some time with Cuban students comparing notes on their University experiences. This was not a tourist experience. The emphasis was on seeking out a more critical and nuanced understanding of Cuba today. In a back street in Havana, students ate and drank in a traditional market, at night they learnt about Cuban dancing or tasted the Cuban nightlife with their University hosts. Everyone left Cuba with an increased insight into this intriguing country, its people and economy. Years later in 2003, I was to read a reflective account by an American woman who had lived in Cuban during the 1990s, entitled *Cuban Letters* (Isadora Tattlin 2002) and found myself recognise that 'experiential geography' we had so successfully captured on our visit to this country.

In 1998, I was 'kicked upstairs' to become Head of the School of Sciences & Sport in which Geography was one of five departments. It did not stop me continuing to teach in Geography at all levels. With Dr David Halsall I taught philosophy and methodology, with Sylvia Woodhead it was countryside management and environment & planning, and with Dr Tasleem Shakur it was post-modern geographies (and anything else he could get me involved with). Post-Modern Geography was my growing passion in these years. In a popular third year course called *Culture, Technology & Space*, we explored the geography of Disney World, The Trafford Centre, *The Simpsons, Wallace & Gromit, Star Trek, Star Wars,* Mobile Phones and the Internet. And along the way engaged with some critical theory from Foucault, Baudrillard and Lefebvre (e.g. Lefebvre 1991, Baudrillard 1988). Was this Geography? Many Geographers have come to define it as a legitimate field for us to study notably Edward Soja who wrote *Postmodern Geography* (1989). A defining feature of this course was the flexibility for each student to choose his or her own final research project for the assessment. This was an individual project given as a report and presentation, which was both tutor and peer assessed. Topics over the years ranged from soap operas to pulp fiction, from zoos to theme parks, from documentary to cartoons, and from Internet dating to video games.

Apart from fieldwork – the first love of most Geographers – one of my most satisfying tasks was supervising a number of third year dissertations each year. I always seemed to have those students with the greatest lack of ideas and the least motivation - except that was until the last year I was at Edge Hill (2002-2003). In that year it all came together for my group of tutees. Their topics perhaps illustrate the 'geography at the edge' flavour of their work: friendship networks and chat room geographies, social spaces of the cartoon *The Simpsons*, realism and the development of the *Spiderman* cartoon (coinciding with the launch of the first *Spiderman* movie and the 9/11 terrorism attack) and the spatiality of *Final Fantasy X*, the videogame. The one thing I remember about geography at Edge Hill was that it was not that stuffy old thing given learned definition in textbooks and dictionaries. It was everything, that is all and anything, which we as geographers found interesting and could apply our geographical way of looking to enlighten a greater understanding. To be human is to dwell, that is to be in a world (to paraphrase the philosopher Martin Heidegger). Or put another way, our identity and our social life are always and already spatial, that is contextual, and to understand it is by definition a geographical question. Next time you phone a friend in Glasgow, or e-mail one in Sydney, just remember the geography.

Today, I lead innovation in learning and teaching at the University of Paisley in Scotland. We do not even have a geography department at our University. Now I work with academics, administrators and technicians to support the continued enhancement of the quality of learning experiences of students across the University. The Centre for Learning & Teaching supports both students and staff, the University's virtual learning environment, providing effective learning skills support, running courses in teaching and learning for staff, and engaging in a myriad of research projects and initiatives. Do I do much geography these days? If geography is what geographers do rather than what they write, then the answer is yes. Now I do not teach geography (a task I miss greatly) but I do occasionally write about the subject (e.g. Rodaway 2004). And what keeps my interest alive in human geography? A continued interest in travel, at home and abroad, near and far, wherever my wanderlust takes me. Most recently this was to Arizona, exploring the landscapes and cultures of the American Southwest. And when I have not the time to pack my rucksack and take to the road, sea or air, I submerge myself in good travel book, especially those which present personal accounts of intimate encounters with different places, different peoples, different worlds from our own. Geography begins with the most familiar, our own body-geographies, but what is ultimately most exciting to me is when our taken-for-granted knowledge of our world and ourselves is challenged by direct experience of other places, meeting the people of other worlds. Whether it is living and working in Romania, a short field course staying with the Berbers in the Atlas mountains of Morocco, or exploring Cuba, for me it is the opportunity to get inside a place and its everyday life. It is opening oneself up to realisation of the relative basis of our own belief and construct of a social and geographical world.

Perhaps the last word should go to 'Che' Guevara, the Argentinian who was famous for his critical role in the Cuba revolution. In his thought provoking reflections on an early journey he made around South America in 1952 (before his later political identity had been fully realised) he demonstrates the transformation from tourist to empathetic outsider or semi-participant. Che bummed his way around the continent, sleeping in guard-houses, hospitals and the homes of ordinary Latin American people from all walks of life, and along the way came to gain a very personal and life-changing insight into the societies through which he travelled. As he wrote of one incident in Peru: 'standing over the small frames of the Indians gathered to see the procession, the blond head of a North American can occasionally be glimpsed, who, with his camera and sports shirt seems to be (and in fact actually is) a correspondent from another world lost amid the isolation of the Inca Empire' (Che Guevara 2004:112-113). This is the kind of 'real' geography we should remember…

References

AREA (1996) Special Issue: *Focus Groups,* including introduction by J Goss (pp113-114) and essays by Goss, Zeigler, Burgess, Jackson, Longhurst.

Baudrillard, J (1988) *America,* Verso: London.

Carpenter, E (1973) *Eskimo Realities,* Holt, Rinehart & Winston, New York.

Guevara, Ernesto 'Che' (2004) *The Motorcycle Diaries: Notes on a Latin American Journey,* Harper Perennial: London.

Hull, J (1990) *Touching the Rock: An Experience of Blindness,* SPCK Books, London.

Lefebvre (1991) *The Production of Space,* Blackwell: Oxford (see also Edward Soja's *Postmodern Geographies* which draws heavily on Lefebvre's work)

Pocock D (1992) 'Catherine Cookson Country,' *Geography 77* pp236-250

Pocock D (1996) 'Place Evocation: The Galilee Chapel in Durham Cathedral,' *Transactions of the Institute of British Geographers* 21.2 pp379-386

Rodaway, P (2004) Two Essays: 'David Ley,' and 'Yi Fu Tuan, in Key' *Thinkers on Space & Place*, Sage Publications, edited by Hubbard, Kitchin & Valentine.

Rodaway P (2003) 'Space, Character & Critique: South Asian Identity in The Simpsons,' in *Picturing South Asian Culture in English*, D'Souza & Shakur (ed) Open House Press, Liverpool.

Rodaway P (1994) *Sensuous Geographies: Body, Sense & Place*, Routledge: London.

Rodaway P (1988) 'Opening environmental experience,' pp50-61 in Pocock D C D (ed) *Humanistic Approaches in Geography*, Occasional Paper, Geography Dept, University of Durham.

Routledge, P (1997) 'The imagery of resistance,' *Transactions of the Institute of British Geographers*, 22 pp359-376

Rowles, G (1976) *Prisons of Space*, Croom Helm: London (an exploration of the geographical experience of a five American senior citizens)

Seamon, D (1979) *A Geography of the Lifeworld*, Croom Helm: London.

Shields, R (1991) *Places on the Margin: Alternative geographies of Modernity*, Routledge: London.

Soja, E (1989) *Postmodern Geographies: The reassertion of space in critical social theory*, Verso: London.

Tattlin, Isadora (2002) *Cuban Diaries*, Bantam Books: London.

Tuan Yi-Fu (1974) *Tophophilia:* Englewood Cliffs, NJ: Prentice Hall.

Tuan Yi Fu (1999) *Who am I? An autobiography of emotion, mind and spirit*, University of Wisconsin Press.

Valentine G (1993) 'Desperately seeking Susan: A Geography of lesbian friendships', *Area* 25.2 pp109-116

Differences in Geography: from Greenhouses to Cotton and from Zimba to Fantasy Football

Nick James[2]

Allotments, the Greenhouse, Seasons and Life itself

It was not this summer but the one before. I remember a TV camera being pointed at me and the journalist asking me why I liked my allotment. What clichés did I express? Perhaps something about being outdoors and being able to exercise as well as grow fresh vegetables. It was not an easy moment. The same angst might emerge if someone asked 'why do you like geography?' There are several reasons for my personal interest in the subject. For instance, I grew up in Africa, I collected stamps, and we travelled lots as a family. I visited countries

---

[2] Nick is currently a Lecturer at the Open University

as different as Finland and Mozambique. Besides those reasons, geography seems to me a subject that has so many corners to it. For example, you can be political or technical, social or scientific, or indeed, all four things at once.

This morning, 14th November, was the first frosty morning of the autumn here in Bristol. I felt compelled to rake the leaves off the grass and then clean out the greenhouse. After a couple of hours work, the garden looked tidier. Then, for a moment, it felt peculiar to be undertaking such an activity. Having grown up in Africa and never having had possession of a greenhouse before, this was all a novel and intriguing experience. It is certainly a rickety old thing and needs attention including wood preserver to give it some strength and protection during the coming winter months. The feeling of weirdness soon wore off and I realised that I must have a healthy interest in both the garden and its greenhouse. Okay, admittedly this is in part to keep on the good side of the landlady, although I have also reached thirty-seven and working in a garden comes as a special pleasure. The niggling question that pestered at the back of my mind, however, was whether this effort was making any positive difference for the ecology in and around our garden.

As you are probably aware, a wide range of opinions exist about how to engage with the environment. The accepted convention is probably to do what I did, and after all, it ends up tidy and therefore looks better. What might a permaculture[3] group suggest? How about organic gardeners? Would leaving it be better for birds and other wildlife? Do the neighbours care?

The greenhouse after all links symbolically to the most pressing and fearsome of global environmental problems, i.e., the 'enhanced greenhouse effect'.[4] Most conventional day-to-day economic and social practices show little evidence of having the capacity to avert the environmental dangers. China's economy is currently growing at an unprecedented rate and therefore demands for oil and other fossil fuels have increased significantly. Here in Britain, car ownership continues to be the pattern, though perhaps once a fortnight we are filling those blue bags or those black boxes with empty cans of lager,

---

[3] 'Permaculture is a philosophy of working with, rather than against nature; of protracted & thoughtful observation rather than protracted & thoughtless action; of looking at systems in all their functions rather than asking only one yield of them & of allowing systems to demonstrate their own evolutions.' Bill Mollison

[4] More commonly referred to as 'Global Warming. The warming of the lower layers of the atmosphere by absorption and reradiation of solar radiation that is thought to increase with increasing atmospheric carbon dioxide

wine bottles, and piles of newspapers in an effort to recycle. Are the global processes of change now moving so fast that we need to think again about and question activities such as these that we think of as normal?

What *is* geography? The discipline is no longer only about learning the facts such as capital cities and rivers, and it has moved on from the 'quantitative revolution' where nearly the whole discipline became subsumed by mathematical models. Phew. Rather, in this article, the narratives touch upon debates about differences that exist over the earth's space, because, as geographers, we have the tools, skills, approaches and the enthusiasm to help towards a fuller understanding of this globalising and fast changing world. Working out and explaining what is going on in particular regions and specific places is important, especially in that quest for a greater understanding.

From the blurb on the back the book *Patterned Ground* (Harrison, Pile and Thrift, 2004) might be an excellent introductory insight to the range and variety of focuses in geography. '*Patterned Ground* is an attempt to unravel the entangled relationships between nature and culture. It consists of about a hundred short entries by some of the leading names in new geography and related disciplines that focus on various 'objects' in the landscape – from beaches to battlefields, bees to horses. Each piece, written by an expert in the field, explores the way in which we understand that object and its relationship to the world around it. This book is neither encyclopaedia nor dictionary, but a knowledgeable and impassioned engagement with the world.'

Cotton cropping and other issues

For my PhD I wanted to learn about the extent and processes of change in the environment as a result of introducing cotton to one area in Zimbabwe. So, the question was whether cotton and its cropping was making the soils less fertile over time. Furthermore, I wanted to investigate the role and impacts of a cotton farming system on household food security. Did the cash-based local economy lead to better provisioning of food through markets and other institutions? My research showed that cotton cropping among smallholder peasant farmer was not enriching for either the people or for the environment. However, getting back to geography:

First, when geography is inter- or trans-disciplinary in its approach, one rarely gets sufficient opportunity to focus in details. I, therefore, for example, went to research in Zimbabwe and worked in a very specific place (see Fig. 9.4).

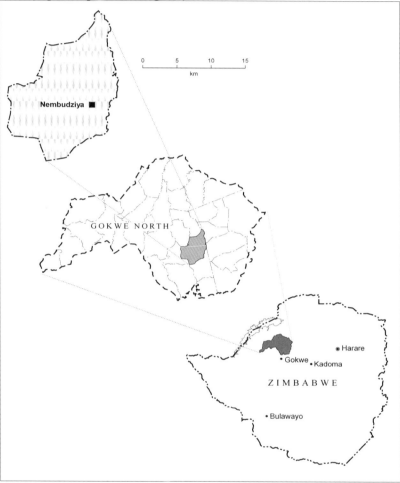

Fig 9.5 Location of Nembudziya, Gokwe North, Zimbabwe, the site of Nick James' PhD research

Map: Ann Chapman

As mentioned above, I researched into food security and environmental change (Fig 9.5). This involved soil fertility analysis in people's fields. However, I can hardly claim to have gone into scientific depth when analysing this (James, 2002). Secondly, by focusing on details from one place one is less aware of the general national and global processes as they unfold and make their impacts on the setting. Therefore, the paradox is that you may be very specific and particularly focused on the one place that you wish to study but the methodological approach to that study will need to be much more general (i.e. mixing different disciplines). Furthermore, that express focus on the one place makes it methodologically difficult to link through to wider processes of change.

I will investigate this methodological quandary by looking briefly at three topics that I addressed while researching in Zimbabwe.

1. Cotton cropping
2. Food security
3. Environmental change

Cotton cropping

Cotton has played a significant part on people's livelihoods in many parts of the world. This is especially so in hot and sunny regions. Cotton is the second export crop in Zimbabwe. It therefore earns a significant amount of income for the country. However, my research showed that it was not such a good crop for the people farming it.

Fig 9.6 Watching the pot: Nick James in Nembudziya, Zimbabwe. Nick read the whole of a book on nutrition while a chicken was cooking on a Mopani wood fire. The luxury of a chicken was a gift. At the time he didn't know the chicken had died of some illness. There is a local (very sensible) taboo against consuming chickens that have died.

Source: Nick James

Cotton directly contributes to land degradation and it increases poverty in many areas. Land degradation refers to the decline in productivity of the soils through over-cultivation, erosion and soil fertility decline. Cotton is an extremely demanding crop, and that in turn can result in over-exploitation of the soil's natural resources. The requirement for cleared fields means less trees, and the hefty amount of weeding means that groundcover is minimal.

That means when it rains, water run-off turns into sheet and rill erosion. Erosion means that the good nutrients in the topsoil may be washed away. Of course, husbandry methods can improve to avert or prevent erosion. However, cotton is such a demanding crop in terms of labour that quite often there is no time left over for soil conservation practice. The technical requirements in the cotton cropping system are therefore contradictory. The heavy weeding needed to protect the seedlings may result in severe damage to the topsoil.

So besides taking up labour time and exposing the soil to erosion cotton demands high levels of the soil's nutrients including nitrogen and phosphorus. With all that time and investment into the crop each smallholder farmer will hope for good returns. However, the major expenses like buying the seed, pesticides and chemical fertilisers eat deeply into potential profit. When it comes to harvest time in April and May there's enormous pressure on farmers. There is so much debt that it's any wonder that there's any left over. If there is it has to go very quickly either to invest in next year's crop or pay for school fees and the like. Making that one cheque last a year is enormous pressure.

Farmers are thus often stuck and trapped into growing cotton. It is the main income earner and the possibilities to diversify are minimal.

In summary, cotton uses up half of the world's pesticides, it is a heavy consumer of soil nutrients (especially phosphorus) it requires clear fields devoid of other crops, weeds and trees and the crop makes an inordinate demand on farmers' labour (James, 2002). My findings and others show that cotton impoverishes regions and increases inequality, thereby worsening social conflict and food insecurity.

Food Security

'Food security includes *nutritional science*; it is about the *politics in the household*, community or region and *access* to enough correct food; it is about the *struggles for livelihood* and the economics of *food availability*; it is about the qualitative constraints of *deprivation, vulnerability, marginality, disempowerment* and loss of *entitlements*' (James, 2002: p. 58).

What this definition shows is both an attempt to grasp and engage with the complexity of the real world but also to use new and perhaps alternative concepts for theoretical understanding. So, yes food security analysis *is* about food shortages, hunger, famine and malnutrition but not exclusively. To be food secure is to remove the *fear* that you will not have sufficient food to eat. Food security also relates to three important shifts in ways of analysing food problems. One is that the focus has moved from a global and national perspective to the household and the individual (Fig 9.6). This gives a more nuanced picture for what may be happening with regard to food security in a particular place or a household. The area I worked in while in Zimbabwe showed hospital records of severe nutritional deficiencies. This was enough evidence to show that a problem did exist in the area. However, the households I worked with were in comparison more food-secure, though some households experienced particular difficulties, and others were vulnerable enough to need food assistance during droughts. The second shift moved from a 'food first' perspective to a livelihood perspective, which encompasses a range of activities an individual engages with to make a living. So the assets, knowledges and capabilities give each individual or household relative 'entitlements' to food and other basic needs. The third shift saw a greater concern with subjective factors as well as the objective assessments. So flexibility, adaptability, diversification and resilience may be important words to think about when analysing food security. Furthermore, it is seen as up to the people themselves to judge whether or not they have reached stressful and fearful situations.

The situation in Zimbabwe is extremely tough and sadly in 2005 getting even tougher for the ordinary people. In cotton regions it is especially difficult for some of the reasons already explained. There are too many other reasons, some of which I address further into the article. The problem of food and access to enough of it is immensely important. Think about this: the very fact that people are food insecure when they need not be is a travesty. All the arguments in BandAid, from the likes of Bob Geldof and Bono, stem from that issue.

Millions of Zimbabweans are today living under conditions of severe food insecurity. The equivocal nature of the concept is perhaps also its strength, in that it can engage with several disciplines and therefore elucidate a more complex set of explanations.

Environmental Change

Ecologies, geographies, the surroundings, and landscapes *change* both through natural processes, *and* through human land use practice. Specifying the environment changes, measuring their rate of change, and explaining what causes the change is critically important.

Fig 9.7 Mixed messages of globalisation in Zimbabwe 1998. A meal of 'eggy' bread, tomato sauce and cornflakes on UN FAO World Food Day. The Zimbabwean slogan on Nick's T shirt Pamberi Ne Zunde Remambo roughly translates as 'forward with the Chief's field system', a reference to an old tradition where the Chief set aside a field, tended by all, for those whose harvests fail. The slogan urges return to a more supportive philosophy rather than the individual profit market-oriented cotton economy.

Source: Nick James

The question of change is thus extremely complex. Firstly, some important questions need addressing. For example, to what extent is that change sustainable? Is the change likely to be manageable and containable without loss of resources or welfare? What is the temporal and spatial extent of the change?

Recent studies have explicitly challenged the assumptions relating to 'crisis' in the African environment (e.g. Leach and Mearns, 1996). It is argued instead that the landscape *transforms* from one state to another. That process therefore need not necessarily equate automatically to a 'crisis' and nor need it be described as land degradation. It is nevertheless subject to serious debate.

What difference does difference make? Africa and its fields

As the world changes, as differences form and reform as the world around us becomes variously more complex and sometimes frustratingly complicated, there remain some certainties. For Sobel (1998: p. 2) 'the latitude and longitude lines... stay fixed as the world changes its configuration underneath them.'

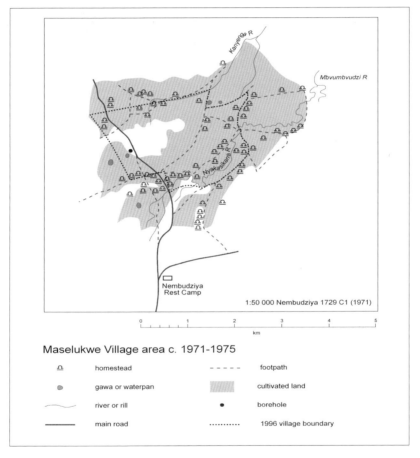

Maselukwe Village area c. 1971-1975

♨	homestead	- - - - -	footpath
●	gawa or waterpan	▨	cultivated land
⌇	river or rill	•	borehole
▬	main road	··········	1996 village boundary

Fig 9.8 Maselukwe village, Zimbabwe, in the 1970s, showing scattered homesteads of mixed farming small holders, a system dating from the 1930s

Map: Ann Chapman

Geography has gained higher respect in recent years mainly because of its abilities to delve wider and deeper towards greater understandings of the multiple complexities that abound among people, places, regions and their countries and continents.

What I aim to do here is to engage briefly with the concept of 'difference' in geographical space. The plan is to illuminate particular points from several scales ranging from continental to the field level.

Africa (within latitudes 40° north and 40° south) is significantly *different* to all the other continents. One of the concerns and puzzles is just how poor Africa is. It is a continent that stands alone in the world. While other parts of the world have witnessed economic growth, increasing trade and technological globalization Africa has remained behind. Since the 1980s the continent has seen falling welfare conditions, severe health problems, and a dramatic fall in life expectancy.[5] However, despite the clichés and stereotypes, Africa is geographically different *within* itself. Zimbabwe and Algeria demonstrate that difference, although of course there are some similarities, especially in the fact that these two nations face deep social and political conflict within their own boundaries.

Zimbabwe is north of South Africa and part of the southern African region. It lies between the Limpopo River and the Zambezi River. It contains both near desert-like dryland conditions as well as near lush tropical conditions with mean annual rainfall of 3,000 millimetres. Compared to the surrounding countries of Zambia, Malawi, Mozambique, South Africa, and Botswana, Zimbabwe is so much better known to us because of the political conflict that abounds within its boundaries.

The land issue is central to the conflict in Zimbabwe. The numbers of white farmers in Zimbabwe fell recently from 4,500 to approximately 400 (BBC Radio 4 2004). Furthermore, a million or more black Zimbabweans have left to form a global diaspora. Since the late 1990s despair, frustration, fear, and deep uncertainty have become part of the situation in Zimbabwe, not to mention the economic crisis (inflation at 400 percent) the desperate food shortages and the HIV/AIDS pandemic.

Within Zimbabwe the analogy of difference continues. Gokwe and the North West of the country are very unlike the rest of the country. It is socio-economically poor; it is the main cotton farming area. The region still hosts tsetse fly[6] and malaria carrying mosquitoes. The area is hot, harsh, and hostile in many ways. The temperature is certainly stifling in October; however, the political tension in this region also has many boiling points.

---

[5] I would urge you to click on to the www.makepovertyhistory.org website
[6] Causing 'sleeping sickness' (*Trypanosomiasis*) in livestock and humans

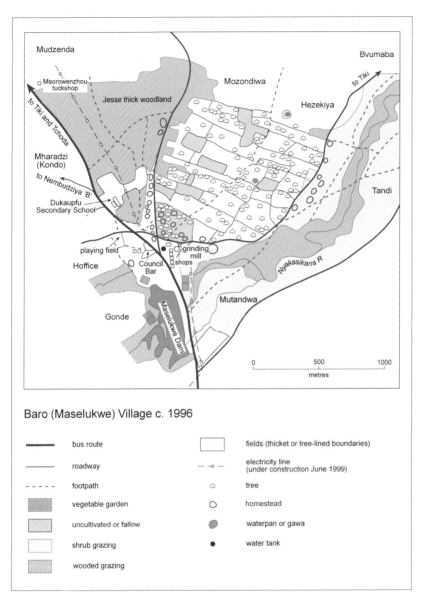

Fig 9.9 Baro, part of Maselukwe village, surveyed by Nick James in 1996, reflecting a 'villagisation' programme, organised along roads, with separate fixed areas of cultivation and grazing.

Map: Ann Chapman

For practical and methodological reasons I came to focus on one Ward within Gokwe, a district in the NW region of Zimbabwe

(James, 2002). Nembudziya is a place some fifty-six kilometres north of Gokwe town situated at 29o00' East of Greenwich and 17o40' south of the Equator (see Fig 9.5).

The switch in analysis from different scales can go on making closer and closer focus. Nembudziya has 10,000 people and over 100 villages. I chose to work in five *different* villages.

Within each village I worked with five different households. Some were relatively poor and at least one in the survey sample was considered wealthy. Within the household space we might here turn to the sociologist or the anthropologist for further analysis. However, it is here that the geographer is also very important. For instance, what is the homestead setting like? How far is it to the field/s? What are the fields like in terms of cropping strategy and soil fertility? How do people make a living?

This analysis gives a sequential scale of representation to demonstrate the existence and relevance of difference in geography. Any analysis of spatial or place changes should accept that complexity.

The 'Zimba Boys' and Fantasy Football

I am deeply interested in place and not just because the concept is seen as central to geography (see Cresswell, 2004). Any cursory look at a dictionary will show the many meanings to such a word. However, don't we all like visiting new places? Whether it is for a holiday or to visit relatives, seeing and experiencing a place offers boundless fascination. I have a very good friend from Liverpool who often teases: pointing to types of building or people he will snort 'How very Bristol.' What can he mean? Is there something uniquely characteristic about the city? Is this a city one imagines full of Swampy-type environmentalists, with dreadlocks and listening to dub reggae?

As you have probably gathered, I live in Bristol. At the beginning of this year's football season the very same Scouse friend of mine persuaded me to join Yahoo's fantasy football competition. I named my fantasy team 'The Zimba Boys'.

I must divert to explain that I spent six years studying for a PhD at Edge Hill College. I started in 1996 with all the reading. I was given some teaching work in the Afro Asian Studies Department. The 5.04am train that began the journey from Bristol to Ormskirk became routine on those ten or so Friday mornings. I enjoyed the seminars. We watched and talked about Billy Connolly visiting Mozambique for Comic Relief. This sometimes became quite emotional: especially the scene where two orphaned children are re-united with their father after a sixteen-year war has ended. Sometimes we had convoluted conversations. In one seminar we were reflecting on the political life of Samora Machel the charismatic former leader of Mozambique. A student asked: 'How can a cattle herder become a president?' I was quick to remind him that many political leaders came from humble beginnings. Ronald Reagan for example was the son of a shoe salesman. I really valued my time at Edge Hill.

Anyway, why the 'Zimba Boys'? Zimba, some fifty miles north of Livingstone in Zambia, is roughly where I lived from the ages of six to ten. The farm was about eight miles south of Zimba, which had a primary school. In the morning, our father drove us to primary school in a Datsun pickup truck. I remember, quite vividly, the classroom. Not all of us had desks. It is very strange to think back; that all those years ago I entered a stereotypical rural African primary school and sat down in a classroom with no glass in the windows.

Very early on into my time in Zimba primary school, we went for a walk into the bush, the local 'wilderness'. The teacher took us out to pick and examine local plants. I cannot remember all the names, in Tonga, of the thirty or so plants that we gathered. I do remember *munego* (*Azanza garckeana*). Its 'common names' include 'tree hibiscus' and 'snot apple'. Another fruit (*Dovyalis sp.*) took me some thirty years to rediscover. It has remained a secret all these years and it is a treat during certain times of the year for livestock herders and other children in rural Africa. That intrigue, that quiet and almost silent knowledge connects and brings together two eras (1970s and 1990s) two parts of my life (childhood and a research student) and two African countries (Zambia and Zimbabwe). The legacy and interest must have remained latent until, in my PhD thesis, I examined the food and medicinal value of some two hundred foods in Nembudziya in NW Zimbabwe (James, 2002).

Well, Zimba is some place that thousands upon thousands of drivers pass through each year including European travellers and tourists. Some stop briefly to refuel or to catch a service stop before the final descent into Livingstone with its Victoria Falls and other tourist trappings. Zimba's front consists of a small row of shops, a bar and a small hospital. It is no doubt an important centre for that particular part of the Southern Province in Zambia.

However, this was where I remember a bicycle running into me. That was scary, although the masked *nyau* (witch-doctor) who chased after me while just about everyone else scarpered was simply petrifying. This was an early introduction to the more mysterious elements of African culture. Everyone from children to adults was mindful, scared and deeply respectful of the *nyaus*. This sub-culture, this underground and mythical/mystical set of people occasionally aggravated the police, but generally, the public avoided and ignored them. The small wood thicket next to the turn off from the main road gave me a deep fear, and no doubt this has lasted into my adult life.

Conclusion

This essay gives a personal insight into my interests in geography. The philosophical argument that 'difference' in geographical space goes on and on poses an intellectual problem. As capitalist development progresses globally, certain processes towards 'sameness' might be observed. Is that a good thing? There is also the geographical problem of place. What can we do to understand such settings and also work towards improving them? Is place the setting from which action and resistance can take place? This article explores these questions and others, and it is hoped that the account helps in provoking a continued interest in geography.

Geography is a very rich subject containing areas of enquiry that go way beyond such concepts of place, space, distance and direction. Geography is undoubtedly an active subject.

References

BBC Radio 4 (2004) 'Voices from Zimbabwe' http://news.bbc.co.uk/1/hi/programmes/crossing_continents/3980 245.stm Accessed on 15th November 2004

Cresswell, T. (2004) *Place. A Short Introduction* London, Blackwell Publishing.

Harrison, S. Pile, S. and Thrift, N. (Eds) (2004) *Patterned Ground. Entanglements of Nature and Culture*, London, Reaktion Books Ltd.

James, N. (2002) 'A geographical study of Nembudziya, Gokwe North, Zimbabwe: The relationship between agrarian environmental change and household food security in a cotton growing area' *Unpublished PhD Thesis,* Lancaster University, UK

Leach, M. and Mearns, R. (Eds.) (1996) *The Lie of the Land. Challenging Received Wisdom on the African environment* London, James Currey.

Sobel, D. (1998) *Longitude. The Story of a Lone Genius who solved the Greatest Scientific Problem of His Time,* London, Fourth Estate Ltd.

From Part Time to Post Doc: what Midges can tell us about Past Climates

Barbara Lang

Beginnings - a New Career?

'No, of course a geography 'A' level from 1976 isn't too old, it's not as old as mine, just tell admissions that it's fine'. These encouraging words from tutor Steve Suggitt in late September 1994 were followed by another phone call and that was it, done. I was now a part-time mature student at Edge Hill studying for a degree in geography. To say I was a shocked was an understatement and I certainly didn't realise how the events of that day would change my life.

I'd left school at eighteen to work in my parents' business and now at thirty-six and the mother of three, I was in business with my husband Andrew. When my youngest daughter started playschool I started to help at my two older children's school: Wesham C.E. primary. I enjoyed this and as the weeks went by I began to think that maybe I could teach too. I finally voiced this thought to a friend. Her reply was 'Go for it, Edge Hill do part-time degrees, here's the phone number, ring now.' Two days later I was in.

I'd always loved geography at school so it was my obvious choice. For the first year I was very part-time, going into College once a week for the Human Geography module. There were also Geoskills sessions and several tutorials, all of these fitted into one morning. The College lifestyle was a culture shock being totally different from anything I'd ever experienced. Since leaving school, most of my activities revolved around my home and family. Now I had gone back to studying and my fellow students were, for the most part, literally half my age. Geography teaching was different and the first couple of sessions with their emphasis on concepts and systems and a reading list that included Paul Rodaway's book *Sensuous Geographies* made me wonder what I'd let myself in for.

The Geoskills sessions were traumatic. In the main sessions it was only the terminology that I struggled with, now I had to deal with computer technology and statistics as well. When the tutor told us to turn the computer on I couldn't even do that. I asked the two young lads on either side of me so many questions that I think they avoided sitting near me for the rest of the year. Despite all this I persevered and by the end of the year I was enjoying learning and developing a hunger for more knowledge.

Choices, some Good Advice and my Second Part-time Year

At the end of my first year I was given a piece of advice that I will never forget. I was taking the physical geography module and a history module; I wanted to drop the history in the second year, however I needed a minor subject that I would continue through my three years of study. Being very sensible and thinking of the future and the subjects that might best help to get a job, I was leaning towards Information Technology (IT). However, an earlier fascination with geology had been rekindled. The problem was that I couldn't see how the subject would help in my search for a primary school teaching job. It was at this point that tutor David Halsall asked which subject I would enjoy most. So far this question had never arisen. I'd spent my life doing what was best or what other people thought I should do, regardless of what I wanted. He really made me think. There was no problem in answering. I knew which I would enjoy more. David's answer was simple; he said that if I enjoyed what I was doing, I'd do much better. His comment actually turned a light on for me. There was no question: it would be geology and for no other reason than I wanted to do it.

Fig 9.10 Barbara's supportive student cohort

Source: Barbara Lang

In my second year I started as part of the student cohort that I would complete my degree with and I made valued and dear friends within this group. My choice of geology was right. I loved the sessions and the use of IT within modules was so extensive, that I was able to continue developing these skills. I gained confidence as the year progressed and became aware of the 'skills issue' realising that although I'd been out of education for several years I had developed many useful skills while bringing up my family, running our shop and just generally living my life. I could organise my time and wasn't afraid of asking if I couldn't understand something.

Divorce, God and Final Two Years

The juggling of a new career and family appeared to be going well. I'd had the support of my husband and friends at home, but our business was showing strains from the recession and we were under pressure. College work was my escape, I loved it and it gave me several hours a week where I could indulge myself in something other than family or business. My husband didn't have this escape however and, the day after my first year results, he left.

The summer of 1996 was a summer of hell for me. After the shock of separation, the distress of breaking up a business and a complete change in home life, I returned to Edge Hill, four stones lighter, to begin Year 2 as a single parent. I returned with a new outlook too. During my summer of trauma, when the bottom had fallen out of my world, I'd found immense consolation and gained strength from my religious faith and, in spite of everything, I was hopeful.

The last two years of my degree were full-time. During this period I had some incredibly low and worrying times. Despite this I can say that I grew more both academically and personally than at any other time in my life. And, despite desperately missing my husband, I actually enjoyed myself and can say that these were probably the two most amazing years I've had. The juggling act of course, got worse. Getting the children to and from school, myself to and from College (which was thirty-five miles from home) fitting in my studies and finding time for my newly-developed social life meant that for two years I never stopped. I already had good and supportive friends at home who picked up and dropped the kids off when my lectures finished late or who had them to stay when I was away on fieldwork. My College friends were also amazing. The group were a mixed bunch, of ages and personalities but all were very special. Throughout they looked after me, made me laugh, laughed at me, listened, bossed me, shouted at me, took me clubbing, dragged me up mountains, helped me move house and decorate it and became an extended family to my children (Fig 9.10).

Being in the second year and full-time meant my workload increased. This kept me going through the ups and downs of divorce.

I would work and read late into the night, catch a few hours sleep and then get up with the kids for school. I studied a mixture of topics including geology, geo-information systems, glaciation and climate and environmental change, which re-sparked a second of my childhood fascinations in climate change. For my third year dissertation I combined several of my interests and looked at the mineralogy of fine-grained glacial sediments from Norway. I learned so much but it never seemed to be enough and I always wanted to know more.

Fieldwork

No account of a geography student's life, or in fact no account of Edge Hill geography would be complete without some mention of fieldwork. Although I'm sure this has always been a part of Edge Hill I consider myself very fortunate to have been in the department when I was. The compulsory fieldwork of the first year to The Lake District involved both physical and human studies. I remember seeing the Lakes in a totally different light after looking at the commercial effects of Wordsworth on Grasmere and appreciating the work of the National Trust much more. On the physical side, heavy rain while we were viewing rejuvenated valleys from the head of Mickleden demonstrated the speed at which over-ground run-off rapidly raises the levels of streams turning them from mere trickles to torrents as they cascade down the side of glaciated valleys. (Nigel Richardson possibly did not plan this demonstration and we almost had to wade back). The first year geology trip took us to Fishguard to see the ancient volcanics and rocks of the St David's area. The second year fieldwork was more exotic with two weeks in Mallorca. During the first week we looked at the geographical aspects of the island, at the geomorphology, the typical Mediterranean vegetation and soils, the effect that millions of tourists have on the already strained water supply and some amazing rillenkarren and other karstic features. We also saw evidence of climate change in massive river deposits and the Quaternary aeolianites and colluvium layers. The second week looked at geological aspects. We visited the geology museum, saw turbidite sequences, the Triassic/Miocene unconformity, a variety of limestones and various evidence that showed the instability of this area in the past. On the final year geology trip to the Outer Hebrides I visited the beautiful and unspoilt Isles of Lewis and Harris, studied their culture and assessed the effects that a super-quarry might have on their pristine environment, visited Skye and saw the famous Quirang (Fig 9.11). Various modules were also complemented by day trips that included visits to Downholland Moss to see sedimentary sequences demonstrating local sea level change and to various quarries and brick works to see the more commercial aspects of geology.

I was able to take part in many other 'Gerry Jaunts' and 'Suggitrips' (as these excursions were fondly called by the students of the day). Gerry Lucas and Steve Suggitt excel in taking students to places where not many students 'have boldly gone' before and students leaving Edge Hill must have some of the most varied field experiences of any university or College. My travels took me to see ancient volcanoes and rocks of Shropshire, the Cretaceous and Jurassic rocks and fossils of the Whitby coast, the igneous intrusions and rock formations of the Isle of Arran, modern day volcanics in Tenerife, glaciers and metamorphic rocks in Norway and desert and gypsum landforms in Almeria.

Fig 9.11 Gerry Lucas being shown how to hotwire a minibus on fieldwork in the Outer Hebrides

Source: Barbara Lang

From a geographical and geological perspective nothing can compare with the experience of actually seeing what you've been learning about in class and in textbooks. To walk on a glacier or a volcano, to see obvious evidence of vast amounts of running water in areas that are now arid and dry, or evidence of higher sea levels, to see pristine beaches or rock formations you've only ever seen diagrams of, to pick fossils off a beach and to measure how fast a stream is flowing - all of this just deepens understanding and awareness, makes head knowledge become heartfelt perception and makes the whole subject real.

From a social perspective the fieldtrips were an experience not to be missed either. I remember Gerry Lucas advising everyone to go on the 'Rocks Make Relief' fieldtrip to Shropshire that was run early in the very first term. He said it was a good way of getting to know everyone. This summed up the social side of fieldwork. Once you'd been on a field trip you certainly knew everyone. I have memories and pictures from these trips that will never be forgotten or matched, here's a few:

- We nearly always got wet as it rained wherever we went, Tenerife, Mallorca, everywhere with the exception of the desert in Almeria. We did have some beautiful weather though and the rain didn't last all the time.
- We always drank. When I started at Edge Hill I got drunk on one glass of wine – by the time I'd finished I could drink a bottle.
- Gerry Lucas always found the highest mountain in the area to climb – he called them hills.
- Steve Suggitt always found roads with incredibly steep drops on either side (especially when I was driving behind).
- The accommodation varied from tents in the middle of the Norwegian nowhere with no running water or toilets (just dig a hole) to possibly flea ridden mattresses in a ranch-type place in the middle of the Almerian desert, with water and electricity that was turned off at midnight by the eccentric English lady who ran the place, to the Whitby boarding house with a never ending amount of floors and only one toilet between them, and where the landlord's brothers-in-law looked like bouncers and smoked flaming joints, to eleven-storey high hotels or apartments in Mallorca and Tenerife, to Youth Hostels (some good, some bad) and finally to wonderful, warm, friendly, clean boarding houses with magnificent views in Arran and Harris where the food was to die for.
- In Norway in September the night skies were so clear and free from light pollution, they were only matched by the beauty of the Almerian skies.
- In Norway in June the sun never set – this confused our body clocks – we went to bed at 3am in the morning,

had breakfast at 11am, walked (and waded) to glaciers by lunch time (4pm) then returned for tea at 10pm before singing and going to bed at 4am.

- Tenerife and Almeria were much more organised. Gerry Lucas had us out at 8.30am – we just managed with less sleep and had coffee breaks at 10am while Gerry went off taking photos of the locals.

- In Almeria our feet burned in our boots the ground was so hot.

- Approaching Tenerife I looked out of the plane window to see nothing but clouds with Mount Teide rising above them. Later that week we looked down on the clouds from Mount Teide – magnificent.

- The experience of being on a glacier and hearing it move will never be forgotten (Fig 9.12).

Fig 9.12 Barbara Lang (right) and Tammy Finlay in front of Bergetsbreen glacier, Norway

Source: Barbara Lang

I was apprehensive of fieldwork at first. I'd never been away from my family and although there were some other mature students in the group, the majority were much younger than me. I was out of condition and found some of the walking strenuous but I was really touched by the camaraderie of the group, both students and staff. As Gerry had said, you got to know everyone on fieldwork but it was more than just getting to know people; group members looked after one another. I'm afraid of heights and we travelled some high roads and scrambled up places I wouldn't have normally gone, but there was always someone there to help and I was never made to feel a nuisance. When my children came to Arran they were included in the same way. I've been on fieldwork with mixed year groups since and realise that this wasn't just specific to my year. The bonding established on field trips continues throughout the degree program, in lessons and free time. In my opinion fieldwork does more for a group and the department than just teach geography.

What Next?

By the beginning of the third year I was applying for a PGCE at Edge Hill. Although still enjoying my time helping at Wesham School and anticipating teaching, the love of my subject had taken over, though the idea of doing a PhD was still a pipe dream. For one thing, although I was heading for a First Class degree, I never thought I was clever enough, and the idea of looking for a post and moving my already disrupted children to another part of the country seemed impossible. When Nigel Richardson mentioned that the department was applying for studentship funding for a PhD in Holocene climate change, I knew instantly that was what I wanted to do.

I continued with the PGCE application and was offered a place here at Edge Hill. The funding for the studentship was not awarded until several months later and as I was not even sure I'd get an interview for this I accepted the PGCE place, feeling quite guilty as friends who'd also applied hadn't got a place and had to look elsewhere. However I did get an interview for the PhD post, a week after my degree results and was offered the position. I was thrilled and apprehensive at the same time.

The initiative for the project to use 'chironomids' from Hawes Water near Silverdale in Lancashire to reconstruct Holocene temperatures had come (so I'm told) from Richard Jones, at that time a PhD student in the department, who was looking through Hawes Water sediment at pollen. Richard Jones had kept finding 'shed loads of these things' in his sediments (Fig 9.13). Richard's fellow PhD student John Hindley had identified the 'things' as chironomid midges. His supervisor, Alan Bedford, was already well known in the department for his love and fascination of these little creatures and was often found wading in ditches and ponds in his search for them. Being up to date with the latest palaeoecological studies, Richard Jones and Nigel Richardson were aware that chironomids were emerging as a new and potentially powerful proxy indicator for

Quaternary environmental change and the setting up of the project was very forward thinking on their part. However neither had any experience of chironomidae. It was here that Alan's expertise came in and the project was set up across the two departments with Alan Bedford and Nigel Richardson as supervisors. One of the benefits of using chironomids was that their abundances in lake sediments could be used to infer past temperatures by means of a computer based transfer function. Steve Brooks from the Natural History Museum in London was very much leading the way in chironomid palaeo-research in Europe at that time and was working on such a transfer function. A visit to the Natural History Museum by Richard and Nigel secured Steve as a third supervisor for the project.

Fig 9.13 Collecting cores in Hawes Water

Source: Barbara Lang

PhD starts

Thus in September 1998 I started my PhD research as a joint student between the two departments. I was based for most of the time in the Natural and Applied Sciences Building in the spacious Advanced Biology Lab. I did feel a bit of a fraud as someone who had only done 'non-exam' biology at school. For the first few months I tried to get my head round chironomid identification. Alan's expertise was invaluable and he was very patient with my terminology. I soon found that identifying sub-fossil chironomids was very similar to identification of real fossils in geology and felt a bit more at home and eventually I did stop seeing them in my sleep.

Though my PhD work fitted in well with my family life shortage of time and lack of finances were my main problems. My finances were slightly eased by various demonstrating and teaching which I was able to do within the geography modules, all useful experience. I was also fortunate in being able to continue with fieldtrips. My time was flexible and I could adapt my work around family life, and once the project got going I was able spend much of the time working at home though I had to be very organised and not let myself get distracted.

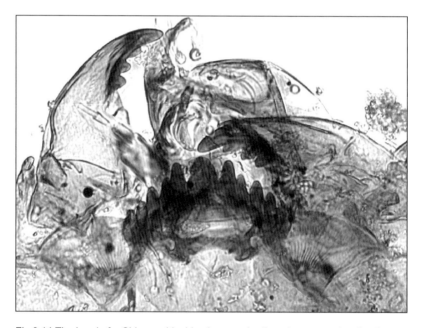

Fig 9.14 The head of a Chironomid midge larva under the microscope showing the many characteristics used for identification.

Source: Barbara Lang

New Methodology

My work itself was a mixture of frustration, tedium and excitement. The literature said that chironomids (Fig 9.14) were found in abundance in sediments with hundreds being found in a few grams. Although Richard Jones had found 'shed loads' in the Late-glacial sediments from Hawes Water this was not the case with the Holocene sediments that were pure carbonates as opposed to clays. Picking chironomid heads from sediments is notoriously tedious but with the sediments from Hawes Water it was impossibly tedious. The standard procedures for chironomid preparation were totally inefficient at breaking up the sediment and thus the search for and establishment of a new methodology started. The successful new method used ultrasound to break up the carbonate sediments and produced largely increased numbers of heads. We then tried the newly established method on other sediments and found that this method increased the numbers of heads retrievable from clay sediments two and sometimes threefold. This was a wonderful achievement as in effect it meant less time was needed to pick more heads from less sediment. I presented the method at the chironomid workshop at Exeter University and the method started to be successfully used by co-workers. My delight in this was soon dampened as we tried to publish the method. I then encountered what I think is the biggest difference between undergraduate work and postgraduate work. Whereas with undergraduate projects all marking is encouraging and positive, I felt a kick in the teeth at the referees' comments of the submitted methodology paper. I'm afraid I would have given up at this point, however Alan Bedford was much more experienced. The main problem seemed to be that we'd dried the sediments before processing. Several tests later and the resubmitted paper showed that the new method was just as effective on wet and dry sediments and that using dried sediments had no effects on total head numbers. The paper was accepted and published but I had become aware of the critical and competitive nature of the postgraduate world.

Core Analysis

The development of the methodology delayed me starting the actual work that had been planned and analysis of the core, which was 8.5m long was still fairly tedious and long winded. However as I began to analyse my results and saw how they were fitting in and complementing other Holocene research I was thrilled. I took my data down to the Natural History Museum in London where Steve Brooks put it through his inference model. The process produced a high-resolution temperature reconstruction for the early to mid Holocene in the north west of England. This data is of both regional and national importance as it is the first quantitative reconstruction for this period for the British Isles. As such it provides invaluable information for a period, which up until now has been little studied and it fills the gap in British research between the much-studied Late

Glacial period and the beginning of many peat records. The chironomid record showed a clear response to the early Holocene cooling event at 8.2ka but also a much larger cooling event at around 9.2ka. Evidence of this 9.2k event was beginning to be seen in other research and relatively new research from the North Atlantic Ocean itself showed evidence for a greater effect in the North Atlantic circulation at 9.2k than 8.2k. The Hawes Water record both confirmed and reinforced suggestions that changes in the strength of the Thermohaline Circulation of the North Atlantic has an important influence on the surrounding lands and the sensitivity of the Hawes Water record to changes in oceanic circulation showed the record to be of vital importance to the current body of research. Agreeing so closely with oceanic records the Hawes Water record may serve as a benchmark for future research. Comparisons of this record, with records from other sites, particularly those situated at increasing distances from the sea, will give an indication of the extent to which oceanic circulation has affected terrestrial climate change in the past. Such knowledge is a vital aspect in the understanding of how future changes in the North Atlantic circulation may affect modern climate systems.

Workshops and Conferences

My confidence in my chironomid skills grew with time as well. Early on in my work I became part of the European chironomid workers group and regularly attended invaluable workshops with this group. One of the problems of working with sub-fossil chironomids is that in most cases only part of the head capsule is preserved and therefore characteristics, which are indicative of particular species, are often missing. This problem has lead chironomid workers to look for additional morphological characteristics, which can be used to identify, sub fossil heads and has also lead to the development of many pseudospecies. Discussion within these groups has been important and has enabled the circulation of information and ideas. It's so easy to think that there's only you having problems with identification of certain taxon or only you having problems with stats. These meetings were so encouraging as they showed me that all co-workers, great and small, had the same problems.

I attended several conferences as a post-grad such as the postgraduate conference at Newcastle University in 1999 and the Quaternary Research Association conference at Southampton in 2000. The highlight however, was the Eighth International Symposium of Palaeolimnology which I attended at Kingston University, Canada in

2000. Here I was able to hear some of the most eminent speakers in the field of palaeolimnology and also be part of a chironomid workshop organised by Ian Walker, America's foremost midge man.

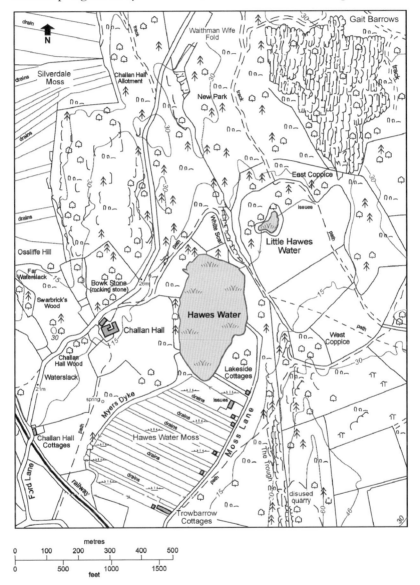

Fig 9.15 Hawes Water and its surrounding limestone area

Map: Ann Chapman

Writing Up

The hardest thing I have ever had to do was write up my PhD I think I would have put it off forever had the offer of a job not come along. Why it was such a problem I'll never know but the thought of a new project and paid employment kept me going. Although bits and pieces had already been written I started in earnest in June 2003. With the prospect of a job work intensified and for several weeks I would be writing from 6am until after midnight. The children were my strength here, feeding themselves and me and supplying endless cups of tea. I'm also indebted to my supervisors Alan Bedford, Nigel Richardson and Steve Suggitt and to Richard Jones, who helped with their comments and expertise. I had finished the basic format by the time my job started in September and I finally submitted my thesis in December 2003. My viva was planned for April, but my father-in law had died two days before it was due and so it had to be postponed until May. I know a lot of people enjoy their viva but I was very nervous and found the whole thing exhausting. I'd finished my corrections and the whole thing was finished by the Christmas graduation. The children had been too young for my first graduation ceremony so I'd gone by myself. This time I got seats for the children and some close friends, so the whole event was marvellous.

Post Doctoral Research

My new job started in September 2004, just ten years after I'd started my degree at Edge Hill. I am now half way through the three-year post-doctorate research project, working in the Earth and Ocean Sciences Department at Liverpool University. My new exciting research builds on both work done by Alan Bedford, Richard Jones and my new boss Jim Marshall. Again I'm looking at midges to identify climate shifts in the North west of England but this study uses evidence for the Late-glacial and early Holocene period from several lakes, not just Hawes Water. The project will combine both isotope analysis and chironomid analysis to infer changes in meteoric water and differences in air mass trajectory. My yearning for knowledge continues and the new project enables me to learn new skills including isotope analysis. However I find I still don't have either enough money or time to do everything I want to do. My links with Edge Hill are still strong not only because Alan Bedford is co PI on my postdoc project but also because it's been such a large part of my life.

My ten years at Edge Hill have not been easy. I suffered marriage break up, financial crises and the death of both of my parents and my father-in law. Through all this I've kept going. The success I see is not mine entirely and I thank God for the love and support I have received from my three children and from my family and friends both at home and at Edge Hill. Without them my achievements would have been impossible. In a world where so many people are retraining and taking up higher education as mature students I would say let the last ten years of my life be a testimony to the fact that it can be done.

Edge Hill Geography Research at Hawes Water

Hawes Water is an important site for geography research by Barbara Lang and others from both Liverpool and Edge Hill. It is the only lake of natural origin remaining in Lancashire. Lying in a lowland limestone area, its twelve metres of calcareous deposits preserve a record of vegetation and climate change through the last 15,000 years. The carbonate sediments are particularly sensitive to climate changes, and also show when people first began to have an impact on the landscape, through the appearance of cereal pollen. Recent work by geographers such as Richard Jones has recorded more climatic oscillations than previously known, reflecting greater climatic instability than previously recorded in North West England, as the climate began to warm up after the last Ice Age. This research helps put into context changes in climate occurring today.

10 Practice makes Perfect

Widespread ownership of computers and access to the World Wide Web have revolutionised experiences of learning and teaching in geography; using the Internet has been incorporated into geographical research as it has into every part of modern life. In this section current geographers explain their research and approaches to geography, while Kathryn Coffey considers what makes up the job of a geography technician. Steve Suggitt luxuriates in the joy of open-air geographical fieldwork research in locations both near and far, while avoiding the 'dark side' of being lured into geology. Nigel Richardson sets much of current Edge Hill geography research into context, explaining why peat bogs are an important environmental archive, recording atmospheric pollution as it falls and preserving past vegetation and even people. Charles Rawding continues the geographical links with education and historical geography, both of which go right back to the early days of Edge Hill in Liverpool. Tas Shakur explains his journeys via rote learning of textbooks in Bangladesh to development geography at Edge Hill, and June Ennis discusses her role as geography administrator.

A View from the Lab

Kathryn Coffey

I had seen the post of Geography Technician advertised in the local paper and as I was a trained Science Technician I was intrigued. What on earth did a Geography Technician do? The only Geography I had been exposed to during my school day was maps, contour lines and capital cities. I made enquiries, applied for the post and was successful. I started working in Geography at Edge Hill in September 1989 and have never looked back. Although I was only employed for six hours a week for thirty-six weeks per year this suited me as it was my first tentative step back to the employment I had trained for since giving up work to have and care for my three children some eight years earlier.

The geography team was quite small and friendly and I immediately felt comfortable in my new surroundings. The laboratory was very compact compared to other labs I had worked in, not what I expected at all. I have since found out that it was originally a store cupboard, albeit a fairly large one. Not much practical laboratory work was undertaken in the department because of lack of appropriate equipment, space and technical expertise. My remit was to change all that. Once word got out that there was a 'proper' technician around more and more was demanded of me. I set about my tasks with enthusiasm and soon had the lab up and running. I enjoyed the challenge of setting up a working laboratory and providing much needed technical support to staff and students. A technician's job is very varied to say the least and this was certainly what I was now experiencing. Not only was I required to work in the lab I was also invited to participate in fieldwork. This meant I was on a steep learning curve, as I had never done fieldwork of this nature before. I hadn't encountered much of the field equipment with most of it being completely new to me. I didn't know an Engineer's Level from an Abney Level but I learnt fast. The time came when I was told I was needed to drive a mini bus. Me? Drive a mini bus? Full of students? It's second nature to me now of course and there is nothing I enjoy more than getting out on fieldwork. Coming back to College tired out from sheer physical graft, sometimes cold, dirty and wet can be very satisfying if the day has been successful and we have achieved what we set out to do. Apart from being out of my usual working environment it's also a good opportunity to really get to know the students.

Over the years I have enjoyed working with different tutors and learning from them. My job has grown tremendously and I am now a full time member of staff and very much part of the geography team. The technical team has doubled in size; there are two of us now. The original laboratory has become the Prep Room (Fig 10.1) and we have another four teaching laboratories plus a Rock Thin Sectioning Laboratory, a Magnetics Laboratory and a dedicated Research Laboratory. Our equipment and resource base has built up steadily over the years. We have a really good geological collection of rocks, minerals and fossils that is continually being added to despite my cries of 'No, please, not more rocks.'

These days a lot of work is done in the laboratory as well as in the field. Laboratory practical sessions are regularly timetabled across the degree programmes and students can also book in to use the laboratory facilities when they have project and dissertation work to carry out. We have a range of analytical equipment that is used when analysing water and sediment samples for both academic research work and student work. I am required to make thin sections for microscopic examination from our own rock collection and although the preparation of thin sections is extremely time consuming it is hugely satisfying. Analytical equipment and techniques are employed regularly and we are continually updating our methods. The students are able to gain valuable 'hands on' experience in a variety of equipment and laboratory procedures equipping them with skills that will stand them in good stead for future employment. My job really keeps me on my toes and my brain cells active.

Fig 10.1 Geography Prep Room

Source: Gerry Lucas

I enjoy my job immensely and working with young people certainly has its advantages - there's always someone who is up for a night out. It's really good to see our young geographers coming into the department for the first time and watching them grow in confidence and ability throughout their three years with us. They leave as educated, well-equipped young adults ready to begin their chosen career. I still find my job challenging and rewarding and working with young people is the icing on the cake.

When I started at Edge Hill I was the youngest in the department. That feeling of youth has long since gone especially as I am now the third longest standing member of staff in Geography – not the third oldest though. I have made many good friends during my time at Edge Hill and the Geography department will always be a special place for me. There are times when I have thought about moving on but then common sense takes over and reminds me what a nice place I work in so I guess I might be here until I retire.

Ice and Rocks: from Britain to Norway

Steve Suggitt

The chill of the wind in your face, the firmness of the rocks beneath your feet, the sweet smell of the countryside, a snatch of bird song, the gurgling sound of running water, being *alive* on the planet, surely that's what geography's all about – at least that's why I became an 'almost geographer', 'almost' because there's always rocks and ice there to seduce you to the dark side. So, where did it all start? Those days in the sixties of walking the gritstone edges of the Dark Peak in Derbyshire, the 'bog-trotting' era, the camping trips in Glencoe and Glen Nevis, the trip to New York and Niagara. Yes I suppose I was doomed from the beginning, that first sand castle near Rhyl, the first ascent of Jack's Rake in Langdale and later the first crampon step on the Gornergletscher, the first smell of sulphur on the top of Mt. Teide. It still carries on, fortunately.

GCE, A-Level and Honours degree were gained and research on glacial sediments in the Isle of Man started, three years studying aspects of the glacigenic deposits of part of the Northern Plain. I'm sure my wife Anne could tell my undergraduates about the joys of fieldwork on the windswept Manx coastline around Shellag Point and The Dog Mills. Sand, silt, clay and pebbles are the only record of what has happened in the recent geological past here. They tell of the succession of glacier advances and retreats, of the rivers and lakes that dotted the landscape here, only to be removed by later ice advances. They tell us about the foundations of current landshapes, of the nature of some of the physical resources present, of the land subsequently modified by people.

Then, reality, the need to find gainful employment, but no need to worry, not yet, one could always try to become a university lecturer. Edge Hill – I knew it wasn't in Liverpool, didn't think it was near Birmingham, but *Ormskirk*, where on Earth was that? One interview later and I was looking at a January 1977 start as a Lecturer in Physical Geography, with a responsibility for teaching Quaternary Studies. In my first year this was to be to fourth year BEd students and the year after to the new BA Geography programme as a third year module. First lecture, on Morphogenetic Regions to the first year BA students, I certainly learnt a lot from that session – still largely ingrained in my memory as 'how not to lecture unless you want to bore the pants off your audience'. Hopefully I'm a bit better now, at least I don't get students trying to fall asleep or hide behind a newspaper – we must have more uncomfortable seats now. So, what's happened in those twenty-eight years, what changes have taken place in Edge Hill's Geography Department since I joined as a raw lecturer?

The Historical Dimension

When I started at Edge Hill, as Peter Cundill left for St. Andrews, I was the only full-time physical geographer, Julia McMorrow (Franklin at the time) who had started some four months earlier was lecturing in both Physical Geography and Afro-Asian Studies. The rest of the department, apart from a young resource management geographer called Nigel Simons, all seemed to be historical geographers, though transport geography and vernacular architecture featured but the department was rather polarised into a narrow field, but things were changing and have slowly evolved into the diverse and wide ranging curriculum we provide today.

In those days the struggles all seemed to be about trying to get more fieldwork into the courses, trying to get some sort of physical geography laboratory (I was offered a portion of a bench in the gardeners' potting shed by my head of department at the time) and trying to develop the courses I and my colleague, Julia, had inherited: interesting times indeed. Fieldwork crept into the second year as well as the first year a couple of years later, physical geography expanded slightly, and a social geographer was appointed to a more diverse human geography area. He was to be my roommate for a few years, a confirmed human/social geographer who never really saw the light, never appreciated the mud, sand, rocks, the water and ice in the way that I do. I remember trying to convert him in Hut 3, Room 4, a wonderful building, best ancient wood that I'm sure had been painted, once. If memory serves me right I think it was sited just to the north of the Gulag Archipelago at the time.

John Cater and I shared this arctic outpost for four years in the early 1980s. I'd just completed writing up about Glacigenic Sediments on the Northern Plain of the Isle of Man for an MSc and John was struggling to keep his PhD going studying racial segregation in Bradford. Competition for dissertation students was strong. I took those willing to work hard, analyse materials and produce some good real scientific results, John took those swayed by thinking that an anthropocentric focus was necessary in geography. Anyway we managed to survive with almost diametrically opposite geographical interests, juggle our use of the room for seeing students, cover for each other when we didn't want to be 'found', and even get some work done.

Fig 10.2 Fieldwork in Northumberland with John Cater (right)

Source: Steve Suggitt

The 1980s saw the development of a skills orientation in the geography programmes. Physical geography laboratory work could be done more easily, technical support was in place to help staff and the move to a more scientific approach to physical geography and eventually environmental science and geology had begun. On the fieldwork front John Cater, Nigel Simons, David Halsall and me pioneered the Gower field course for our second year undergraduates. A residential week in Mumbles saw the Welsh valleys, the Lower Swansea Valley, The Gower Peninsula, Rhossili and Swansea the sites for our varied human/physical/resource attack on the geography of the region. The days in field, sun, wind, rain, we had it all, the evenings getting the insides wetter than the outsides, waiting for 'whose round is it next?' to be played out. Then, in the late 1980s, a change for the second year undergraduates, John and I went to the windy north east, Seahouses in Northumberland, and developed an integrated physical and human field week there (Fig 10.2) visiting the Farne Islands, Bamburgh, Craster, the Cheviots, Lindisfarne, urban farms and the Byker Wall. The late Paul Gamble and I formed a successful partnership for a couple of years after John was pulled away by duties in Urban Policy and Race Relations and then the venue survived a metamorphism into a purely physical geography trip, with

Martin O'Hanlon, and eventually a geology trip for the BSc (QTS) Environmental Science degree, with Gerry Lucas.

In parallel with this in the early 1980s the third year geographers were treated to the scenery of Central Wales by Julia McMorrow and me as we studied Quaternary glacial and periglacial features, from solifluction lobes to pingoes, from lodgement till to raised beach deposits.

This was the time when we got real laboratory facilities in the department, thanks to our new Head of Department, Derek Mottershead, a third physical geographer to add to Julia, now physical geography and remote sensing, and myself. The remote sensing laboratory followed and Julia and I moved into adjacent rooms converted at that time in the main geography block. John was left on his own, still a confirmed social geographer and still in Hut 3, Room 4 until his move to the Household Corridor and more work in Urban Policy and Race Relations.

As time moved on we gained a second laboratory for small group teaching, staff and undergraduate research, still in use today. Derek Mottershead and I pioneered foreign fieldwork for our second year undergraduates, to Mallorca. What a different experience for the students: young rocks and landforms in a different environment to the British Isles. The venue has been developed over the years, Nigel Richardson, Sharon Gedye and Annie Worsley all contributing along with me and recently Annie and I have used the area successfully for dissertation and project training, encouraging the students to develop the field skills and fieldwork ethos that underpin all geosciences.

The 1990s saw major developments in our programmes. I developed and headed a new Earth and Environmental Science BSc and the Geological Science minor subject, probably the only new geology section in higher education in the last twenty years, and largely thanks to the appointment of my co-geologist Gerry Lucas, one might say a split persona between his main love, geology, and his work in remote sensing and GIS. Thanks to Gerry (replacing Julia when she moved to University of Manchester) it became possible to develop geology as a separate discipline at Edge Hill. I'd taught introductory and resource geology within the BSc Geography programme for several years as well as tutoring part time in it for the Open University since 1979. With Geology established I was able to move on to developing a Physical Geography and Geology degree, giving the department a three pronged focus, particularly useful with the demise of geography as a discipline in secondary schools over recent years, but the wheel of time will turn again, of that I'm sure with the range of skills and synthesis geography has to offer. What other subject integrates so many areas together and allows us to study and try to understand the changing relationships between people and the planet on which they live?

Geology brought with it a whole set of new challenges. Our rock collection has grown over the years, often to the chagrin of our Senior Lab Technician, Kathy Coffey. I can hear the plea now, 'Don't bring back any more rocks, we haven't got anywhere to put them,' but, as ever, it goes unheard by Gerry and myself; the last samples of Coquina sand from the Dee Estuary and Carboniferous limestone from Halkyn Mountain were collected only last week. Microscopes and thin-sectioning equipment purchases, training course in thin sectioning done in Glasgow and we've not looked back.

Fieldwork for the first year geology students started in Pembroke; Goodwick Volcanics and Strumble Head pillow lavas then moved to Bude in Cornwall after a couple of years, Cornish granites, Delabole Slates, Tintagel Volcanic Formation, Crackington Formation, Bude Formation with its fish coprolites and xiphosurid trails. Three years there and on to Arran, where we delight in one of Hutton's famous unconformity sites, desert sandstones, and coastal dyke swarms, and the obligatory visit to the newest distillery in Scotland at Lochranza for a dram of the light and flavoursome Arran Malt. This year we've moved again, focussing more strongly on field mapping skills in Ardnamurchan. As for the second year students a varying programme of younger, Mesozoic and Cenozoic, sediments in Mallorca, the Banyalbufur Challenge, detailed graphic logging at Estellenchs, palaeoenvironmental interpretations at Costa de la Calme, the Miocene trace fossils of Cala Portals Vells. The third years have focussed on resource geology, mineral planning and the extractive industries more, thanks to Gerry's knowledge and experience in these areas, and have an enjoyable experience in the Western Isles. Some of the oldest rocks in Britain are visited here, the Butt of Lewis, the South Harris Mountains, the fabulous Machair sands, south to the Uists and Benbecula and home via the Isle of Skye and the Cuillins.

Consolidation over the last few years under the leadership first of Paul Rodaway and recently of Nigel Richardson has seen our portfolio of teaching and learning strategies gain recognition from the Geography, Environmental and Earth Sciences (GEES) Subject Centre with a number of staff having publications in this area. The department is now merged with our biologist colleagues to form the Department of Natural, Geographical and Applied Sciences enabling us to work more closely with biological scientists in the future.

Ice and Rock

I see field experience as the essential characteristic of the geo-scientist. It is often quoted that the best geologist is the one who's seen the most rocks, hopefully in the *field* not just the laboratory. Only through a range of field experiences can we begin to appreciate the natural world, its complexity, often its fragility and how an ever changing range of natural processes inexorably bring change to our planet.

So how do we foster this belief in our students? To me the answer is simplicity itself: let them experience it. One of the main successes I have had at Edge Hill is to bring in the concept of enrichment fieldtrips. I feel that students appreciate the willingness of staff to run additional, enrichment fieldtrips.

Fig 10.3 Studying glaciers in Norway

Source: Steve Suggitt

In 1988 I took my first set of students to Norway as part of a joint venture with Lancashire Polytechnic Expedition Society, their second trip, our first. Was it a success? Well I've been taking groups back every year since, seventeen visits in all (Fig 10.3). I now even get Christmas cards from one of the campsites we use at Gjerde. Student feedback is excellent; we camp, often not on campsites with the only facility being running water from snowmelt streams, and the odd boulder. What have they liked best? Was it a June/July trip being able to sleep outside in the near daylight, easily reading at two o'clock in the morning, or was it a September/October venture with the Milky Way clearly visible behind the occasional shooting star overhead. The record stands at four consecutive visits by one student, every year as an undergraduate and when he was doing his PGCE year. Norway and I have had a love affair since my very first visit; countless dissertations achieved and miles walked up Austerdalsbreen or Bergsetbreen or maybe Nigardsbreen. The deep glaciated valleys, a kilometre wide, a kilometre high, the glacier in the distance as we approach, the moraines, boulder fields and sandar we cross to get there, the view back down valley from on the glacier itself, the cool of the ice, its blue colour in ice caves, the gurgle of meltwater down a moulin, the crossing of crevasses, the sun, the wind, the blue sky (if you're lucky) what could make a more lasting impression?

Research has covered topics from glacial meltwater chemistry, morphology and sedimentology of glacigenic landforms and their materials to sourcing, mineralogy and magnetic properties of tills and other glacigenic materials, from asymmetry of vegetation distribution with aspect in a glacial valley, acid deposition on vegetation and palynology to lichenometry and Holocene environmental change. After being a regular visitor since 1988 I've managed to amass enough material to start looking in detail at the sourcing and nature of the sediments being produced by a variety of processes in the area focussing first on the upper part of the Krundalen valley by Bergsetbreen. The samples are collected, now all I need is the time to work on their mineralogy, textural and magnetic properties.

Then, there's Tenerife, a volcanic peak rising from the Atlantic floor, subtropical vegetation and climate, young Neogene and Quaternary basic and intermediate volcanic rocks as opposed to the 2000 million year plus Pre-Cambrian gneisses of the Jostedal in Norway. To trek up to the summit of Mount Teide, permit in hand, to take in the view, to cough at the sulphur fumes and then to descend to the caldera floor, well I'm sure all those students who've been to Tenerife with Gerry and me will have just as lasting an impression. Almeria and Santorini with Gerry, China and Morocco with Tas, these are all experiences that beg you to engage with nature, with the world outside the lecture theatre, these are what foster the inquiring mind. This is my geography; this is where my life's been, I hope you're able to enjoy the same.

Raised Peat Bogs as Environmental Archives

Nigel Richardson

Introduction

Raised peat bogs are wonderful places for undergraduate fieldwork. The relatively low variety of plant species on the peat bog surface is a good starting point for investigating spatial variations in vegetation communities and the environmental factors that may influence the distribution of vegetation. Beneath the surface, the accumulating peat retains a rich and unparalleled archive of our past, as pollen, dust, chemicals (and sometimes human bodies) have fallen onto the bog surface and become trapped and preserved. For the past eleven years geography students from Edge Hill have worked on some of the peat bogs of the south-east Lake District (namely Heathwaite Moss; Helsington Moss; and Fish House Moss) as part of residential fieldwork (Fig 10.4). Some have been inspired by this first encounter with the wet, brown peat and what it can tell us about the past, to complete dissertations on topics including the study of vegetation history, atmospheric particulate pollution histories, and the reconstruction of past climates. In this account I will review some of the more recent uses of raised peat bogs as archives of our past – 'the living history book'.

What is a peat bog?

Peatlands develop in waterlogged conditions where the rates of plant breakdown are low because of a lack of oxygen and acidic conditions. The result is an accumulation of organic matter – peat. The focus of this chapter is on the records from a particular type of peat bog, the raised or **ombrotrophic** ('rain fed') bog. These are dome-shaped accumulations of peat that have grown upwards above the mineral groundwater table, and so are entirely dependent on the atmosphere for all their moisture and nutrients.

Peat bogs are very wet environments, as many of our past students and staff will confirm, a careless footstep plunging them knee-deep into cold water. Up to 98% of a raised bog is water and only 2% solid peat. Each year of the field course we get the students to jump up and down on the bog surface to experience the unstable nature of the material below their feet – a soft living carpet which floats on material that is nearly all water.

Fig 10.4 Edge Hill students taking a peat core at Fish House Moss, Cumbria

Source: Annie Worsley

Generally the peat bog water table is only a few centimetres below the surface. The surface of the bog typically consists of low hummocks up to 50-80cm high, interspersed with hollows and pools, which may be filled with water. This creates zones of characteristic vegetation, each depending on their proximity to the water table. *Sphagnum* moss (the bog moss) is the main component of raised bogs and there may be as many as ten different species, showing great variation in colour from green and yellow through to red and brown. The drier hummocks are often topped with heather (*Calluna vulgaris*) and cotton grass (*Eriophorum vaginatum*) whereas in the wetter parts of the bog, white beaked sedge (*Rhynchospora alba*) and the yellow flowered bog asphodel (*Narthecium ossifragum*) grow.

Beneath the surface, raised bog consists of two layers: the upper layer, where vegetation can decompose, is called the **acrotelm**.

It can vary in depth from 10cm to 50cm and is relatively oxygenated and through which water can flow rapidly.

Below this layer is the **catotelm**, which may be several metres thick. It is a waterlogged and acidic layer that contains very little oxygen and, therefore, vegetation decomposition is very slow. The rate of decay of vegetation slows dramatically as plants pass from the acrotelm to the catotelm.

Pollution records in peat

Since ombrotrophic peat bogs depend entirely on the atmosphere for their nourishment as they accumulate, they often preserve a record of particle deposition from the atmosphere. *Sphagnum* moss is particularly efficient at trapping particles. Some of the more easily measurable types of atmospheric particulates incorporated in peat are magnetic spherules and heavy metals such as aluminium, copper, iron, nickel, lead and zinc. These arise from industrial and urban processes such as solid fuel based power generation, iron and steel manufacture, and non-ferrous metal smelting. Therefore, the magnetic mineral and heavy metal record in ombrotrophic peat can potentially be used for reconstructing atmospheric particulate pollution history. One key criterion that has to be satisfied if peat is to be used as a pollution monitor is that the substances (in this case magnetic particles and heavy metals) do not undergo any changes following deposition onto the peat bog surface.

High concentrations of magnetic atmospheric particulates rich in iron oxides such as magnetite and haematite are produced as a result of fossil fuel combustion. Magnetic measurements of peats from a range of sites throughout the British Isles suggest that the concentration of magnetic particles increased by up to three orders of magnitude shortly after the industrial revolution until the present day, even at remote sites (e.g. Oldfield *et al.* 1978; Richardson, 1986; Clymo *et al.* 1990). However, serious doubts have arisen over the persistence of the magnetic record in ombrotrophic peat. The chemically reducing conditions prevalent in the waterlogged peat of the catotelm layer, together with the high acidity of the bog water would seem to favour a degree of partial dissolution of the magnetic particles in the peat. This appears to distort rather than entirely destroy the depositional record of atmospheric particles.

Heavy metals deposited into peat have their own difficulties associated with uptake by plants, the decomposition of the peat, chemical changes in the peat, and diffusion and downwash in the bog water. Many studies completed in the U.K., Europe and North

America have evaluated the persistence of heavy metal records in peat. In general, it appears that problems of relocation of metals occur mainly in the acrotelm peat. Metals that are either deposited directly into waterlogged peat (i.e. directly into hollows or pools) or pass into the waterlogged catotelm layer appear to be less mobile. This cannot be assumed for all metals. Differences exist in the behaviour of each metal within the peat column, with lead and copper providing some of the more reliable results (Clymo *et al.* 1990; Jones, 1997; Martinez Cortizas *et al,* 2002).

Therefore, in favourable conditions, raised peat bogs can record at least qualitative trends in the past deposition of atmospheric pollutants.

Palaeoclimate records from raised peat bogs

Raised peat bogs contain a detailed terrestrial record of past climate often going back in time up to 5000 years ago. Various approaches have been used to decipher this record.

As noted above, moisture status is a major control on the composition of vegetation communities on raised peat bog surfaces. These vegetation types contribute to the formation of the peat, and their remains can normally be identified as macrofossils within the peat column. Variations in the nature and degree of preservation of such plant macrofossil remains with increasing depth beneath the bog surface reflect changes in the moisture status of the peat bog surface (known as bog surface wetness). Since surface water on ombrotrophic peat bogs is solely derived from precipitation, these data form the basis for high precision palaeoclimatic reconstruction (e.g. Barber *et al.* 2003).

A second approach involves the use of a group of organisms called testate amoebae. These are protozoa that are sensitive to changes in the moisture status of the peat bog surface. After death, their shells are preserved in the peat and can be extracted, identified and counted to allow a reconstruction to be made of changes in surface wetness on peat bog surfaces (e.g. Charman *et al.* 1999).

Finally, peat surface wetness also influences the amount of plant decomposition that takes place in the surface layers of peat (the acrotelm). A drier peat surface is associated with a deep acrotelm, meaning that the upper peat layer is aerated and increased decay of plant material can take place. Conversely, when the surface is wetter, the acrotelm is very shallow and plant remains pass rapidly into the permanently saturated catotelm, where decay is much reduced. Variations in the degree of decay ('peat humification') with depth down a peat column can be interpreted as records of changing bog surface wetness. We currently get our first year students to make a subjective estimate of peat humification on a 5m peat core spanning approximately the last 5000 years from Fish House Moss, Cumbria during the field course. More sophisticated and accurate laboratory techniques are available (e.g. Langdon *et al.* 2003).

Gareth Thompson, a PhD research student in the Geography department at Edge Hill between 1997 and 2000, used all three of these approaches on peat cores obtained from sites in the south Lake District.

Conclusion

Raised peat bogs can be found around the world, offering an opportunity to study the history of atmospheric pollution and past climates on a global scale, as well as providing a valuable educational resource. Unfortunately, these peatlands are being destroyed for agriculture, forestry, urban development, mining of peat for power generation, and for horticultural uses. It is vital our remaining raised bogs are preserved as environmental archives for future generations.

References

Barber, K.E., Chambers, F.M. and Maddy, D. (2003) 'Holocene palaeoclimates from peat stratigraphy: macrofossil proxy climate records from three oceanic raised bogs in England and Ireland.' *Quaternary Science Reviews* 22, 521-539.

Charman, D.J., Hendon, D. and Packman, S. (1999) 'Multiproxy surface wetness records from replicate cores on an ombrotrophic mire: implications for Holocene palaeoclimate records.' *Journal of Quaternary Science* 14, 451-463.

Clymo, R.S., Oldfield, F., Appleby, P.G., Pearson, G.W., Ratnesar, P. and Richardson, N. (1990) 'The record of atmospheric deposition on a rainwater-dependent peatland.' *Phil. Trans. R. Soc. Lond.* B 327, 331-338.

Jones, J.M. (1997) 'Pollution records in peat: an appraisal.' In *Conserving Peatlands* (Parkyn, L., Stoneman, R.E. and Ingram, H.A.P, eds) pp. 88-92. Wallingford: CAB International.

Langdon, P.G., Barber, K.E. and Hughes, P.D.M. (2003) 'A 7500-year peat-based palaeoclimatic reconstruction and evidence for an 1100-year cyclicity in bog surface wetness from Temple Hill Moss, Pentland Hills, southeast Scotland.' *Quaternary Science Reviews* 22, 259-274.

Martinez Cortizas, A., Garcia-Rodeja, E. and Weiss D. (2002) 'Peat bog archives of atmospheric metal depostion.' *The Science of the Total Environment* 292, 1-5.

Oldfield, F. Thompson, R. and Barber, K.E. (1978) 'Changing atmospheric fallout of magnetic particles recorded in recent ombrotrophic peat sections.' *Science* 199, 679-680.

Richardson, N. (1986) 'The mineral magnetic record in recent ombrotrophic peat synchronised by fine resolution pollen analysis.' *Physics of the Earth and Planetary Interiors* 42, 48-56.

From Lincolnshire to Lancashire

Charles Rawding

I came to Edge Hill in September 1997 after a long period teaching geography in a very large 11-18 comprehensive in Humberside. The PGCE Geography was in its infancy and I was the first geographer from the Geography Department to be involved, the previous year having been run for a small course on a part-time basis by a local teacher.

My geographical life began in earnest at the University of Sussex where my first degree was BA (Hons) Geography in the School of Cultural and Community Studies (1976-1979). For me the attraction of Sussex lay in the inter-disciplinary nature of the courses it offered, and while my Geography tended to specialise in social and cultural aspects of the subject, I also took courses on 'Popular Culture, Leisure and the Social Order' (including a memorable fieldtrip which effectively was an (historical) pub crawl) urban history and developmental psychology. This was followed by a PGCE at Sussex in Social Studies (Geography was not on offer and the way forward appeared to be towards Humanities in the years before the National Curriculum). A career in secondary teaching followed.

During my time in Lincolnshire I developed my interests in local history and historical geography, completing an MA (1984) and a DPhil - *A study of place: the North Lincolnshire Wolds 1831-1881* (1989) part-time at the University of Sussex with Professor Brian Short as my supervisor. During this time I acquired experience in HE as a tutor for the University of Hull Dept of Extra Mural Studies and for the Workers Educational Association (WEA) publishing a number of village histories and articles on nineteenth century rural Lincolnshire. In some ways, this period of research was my most rewarding, investigating the area in which I lived, teaching about my findings to adult classes and publishing those findings through the WEA. In addition to these village histories, I produced a number of articles spinning off from my MA and DPhil research. In 2001 (Rawding 2001b) the DPhil was transformed into a book covering the whole of the Lincolnshire Wolds rather than the narrower focus of the original study (Fig 10.5).

Since 1997 most of my work has been in geographical education running the PGCE course which has now trained over 150

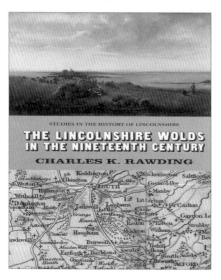

Fig 10.5 Cover of Charles Rawding's historical book

geography teachers including a significant number of Edge Hill graduates. The course has gradually built up a reputation in the north west, and one of the more rewarding aspects for me now is having mentors in place that did the course themselves. As part of my work in geographical education, I have continued my earlier work producing a range of articles principally for *Teaching Geography* (Rawding 1998 for example) and most recently including joint pieces with former trainees on the PGCE course.

For me, one of the exciting aspects of moving to Edge Hill was the opportunity to combine my experience as a classroom teacher with the possibility of working with undergraduates while continuing my own academic research. I am module leader for Modern Historical Geographies, a module first introduced in 2000 reflecting my own research interests in the area and also for Global Tourism and Citizenship. The move from Lincolnshire to Lancashire has also enabled me to shift my research focus towards the northwest of England. I was delighted to be able to resume working with Brian Short as a local researcher for his national project on farming in the Second World War. This work has so far resulted in one published paper and a number of very interesting responses from local groups regarding their experiences during the war. My third hat, that of tourism geographer, generated a paper on Amsterdam as a spin-off from the Second Year undergraduate field trip to Amsterdam (Rawding 2000) and a contribution to a book on sustainable tourism in the Lake District

(Blakey et al 2004). I am also the Secretary of the Central Lancashire branch of the Geographical Association, organising a lecture series, occasional conferences and field visits.

Selected References

Blakey P A, Hind D W G, & Rawding, C. (2004) 'The Lake District as a tourist destination in the 21st century.' In Hind D W G & Mitchell J P (Eds) *Sustainable tourism in the Lake District.* Business Education, Sunderland.

Rawding, C. (2001/2002) 'Lincolnshire aristocratic landscapes : the view from the 1790s.' *Lincolnshire Past and Present*, 46, Winter, p7-8.

Rawding, C. (2001b) *The Lincolnshire Wolds in the nineteenth century.* History of Lincolnshire Committee.

Rawding, C. (2001a) 'The virtual world of tourist cities.' *Global Built Environment Review* Vol 1 No2. p10-12.

Rawding, C. (2000) 'Tourism in Amsterdam: marketing and reality.' *Geography*. 85.2. p 167-172.

Rawding, C. 1998 'The implications of the new Standards for Qualified Teacher status.' *Teaching Geography*. 23. p 40-1.

Transforming Bangladeshi Tas

Tasleem Shakur

Growing up in what was then *East Pakistan*, my first contact with serious 'Geography' was in a 'Sandhurst style' semi-military boarding Cadet College during my Secondary School Certificate preparatory period (1966-1968) near a sleepy hollow of *North Bengal* on the bank of the mighty *Padma* (Ganges) river. Being a long time British Colony, the newly independent state of *Pakistan* was then proudly introducing British public school type boarding institutions strictly following English school curriculum. So at the age of fourteen my Geography textbooks were Monkhouse's *Principles of Physical Geography* (1965) and Dudley Stamp's *Regional Geography* (1960).[1] Living in a tropical region neither text seemed to me to relate much to my surroundings and the only option left was to memorise as much as possible, which guaranteed a high mark. I still remember the long definitions for Igneous Rocks, Plains and Volcanoes (with some abstract dark schematic drawings) for physical geography and detailed descriptions

---

1 Ironically Sir L D Stamp was the first professor of Geology at Rangoon, Burma, where the need for good textbooks was borne in on him. He produced *The World*, a cheap, well-illustrated text, written from a South Asian viewpoint (White 1968) SW

of Iraq's Tigris and Euphrates rivers (with excellent illustrations through historical/vegetational maps of the Mesopotamian region) and its thriving agriculture base in regional geography (ironically today we only see the devastation of the country following the controversial war). While I can't say that I enjoyed geography very much, it was at this institution that I heard much of English pop songs.

My other notable achievement at the Cadet College was a serious involvement with English plays. Much of the credit should go to Mr Hugh K Lambert, an enthusiastic and dedicated Peace Corps volunteer from Colchester. With instructions coming from Mr Lambert's mother living in England, we performed *The Merchant of Venice* with myself (and I am going to say this pompously as this is one of my career's high points) earning the best actor's gold medal for my characterisation of Shylock. Sometimes I look back and think whether it was my early contact with Englishness which has now pushed me rather intriguingly towards 'hybridity' and cultural geography (Shakur and D'Souza 2003).

Influenced by the hippie movement of the West, I chose architecture for my higher studies at the country's only Engineering University at Dacca. However, before we could start our course in came the Pakistan Army crackdown of 25 March 1971 and the nine months of fighting by the oppressed Bengali freedom fighters. Being a Bengali teenager I was greatly shocked by army atrocities, which I personally witnessed. Although I did not join the guerrilla warfare (as some of my peer group did) I was one of the first students to boycott University and I provided mental support to other fighting guerrilla students who fled to India. One of the best memories of my late teen age was that of 16 December 1971 when Pakistan Army surrendered giving birth to a newly born nation of Bangladesh. In the excitement I drove a Volkswagen van with my younger brothers and friends to the city centre only to come face-to-face with the retreating Pakistan Army. As they pointed their rifles I managed to do a quick U turn and speed away.

After indoctrination to Englishness in my early childhood in Muslim majority Pakistan, I was then thoroughly Americanised through Texas AM University-led Architecture education in secular Bangladesh. Obtaining a first class degree I knew more about Sears and Twin Towers than the huts of Bangladesh, which was an odd introduction to a future in development Geography.

Obtaining a UN fellowship in 1977 to pursue an MA in Urban and Regional Planning also finalised my relatively early marriage as my parents wouldn't take any chance of a western-leaning young person like me who may 'tie the knot' to an English woman. Having heard so much of England (much of it came from travel stories by pro-colonialist Indian writers like Nirod Chowdhury) coming to UK was a sort of anti climax for me. Having seen Lutyens' New Delhi, with its spacious vistas, roads and lofty buildings, Piccadilly Circus seemed a narrow lane in Old Dhaka. Having said that 1977-1978 is still one of my memorable years. Travelling from St Pancras to Sheffield on an August Sunday in 1978 was an awesome experience. The English countryside looked so lush green it destroyed my pride in Golden Bengal or even the natural beauty of Kashmir. Taught by the geographer Professor Jimmy James (who we were told was also a Queen's advisor for Town Planning) with a mixture of left-wing young tutors (reflecting the trade union movement of the time) the MA changed my views and politics. I enjoyed the illuminating field trips to the slums of the Gorbals (Glasgow) the Mile Long Housing (Sheffield) Ebenezer Howard's Green Belt or the New Town of Milton Keynes. Returning back to Bangladesh, my MA dissertation was focussed on the upgrading of one of the poorest slums in Dhaka. With acquired radical thoughts from the north of England (particularly those obtained from the socialist labour Councillors of Sheffield) I found it rather difficult to cope with the reactionary physical planning and urban poor housing programmes where I was involved as a team member with the UNDP and World Bank (National Physical Planning with the Urban Development Directorate of Bangladesh). With increased interest in Urban Poor Studies, when I was offered a Commonwealth Scholarship for a PhD study on the squatters of Bangladesh at Liverpool University, it came as a great relief to escape the frustrating conditions in Bangladesh.

During 1982 I chose Liverpool to be close to a perceived Beatles identity. Once again my naivety did not prepare me for the recently riot-torn city. Living behind the Philharmonic Hall, in the University's married students flats, I considered Liverpool 8 to be a safe place to live (particularly considering I was only mugged once). I also witnessed for the first time the multiple deprivation of a developed city like Liverpool. Working as a part time waiter near Allerton I found young boys of ten to twelve years approaching the 'curry houses' for washing up jobs. During those Margaret Thatcher days I saw the radicalisation of Liverpool Council under militant leader Derek Hatton who once gave a lunchtime illuminating talk to the 'Dept of Civic Design' student society with a pint of lager. During this period I saw a huge area near the Roman Catholic cathedral getting blighted with empty houses. It was a great time to understand urban deprivation, and I still take students to see urban revival near China Town and the regenerated Albert Dock area.

Fig 10.6 Tas in China

Source: Tasleem Shakur

After the completion of my doctoral research at Liverpool I sent my family to Bangladesh preparing myself to go back to my country. However, that was not my destiny as I was offered a post doc Fellowship at Oxford Polytechnic (now Oxford Brooks) to work on an ODA (Overseas Development Agency, now DFID) funded international desk research on Unregulated Urban Housing Submarkets (Shakur et al 2001). Before finishing my term at Oxford came a three-year research position at my old University of Sheffield (Dept of Town and Regional Planning). Being away from home and family was not easy hence I had to recall my family to Sheffield. While in my second year contract came my first lectureship at SOAS (School of Oriental and African Studies, University of London). It was a fixed term lectureship in human geography with particular reference to South Asia to replace Professor Graham Chapman (now at Lancaster University) who was setting up an International Development Centre within Geography. This was a turning point in my career as it was the first time I formally started to teach in a Geography department. It was an exciting two years at SOAS but I was getting very exhausted as I lived weekdays in a Camden council flat (subletting from a Bangladeshi catering person) and meeting my family in Sheffield over the weekend. To top up my meagre salary I started doing three other part time lectureships at Kingston Poly (Geography) Thames Valley

Poly (Land Survey) and Sheffield University (Landscape Architecture). On top I even managed to mark Open University (D205 Changing World, Changing Britain) scripts during the intercity journeys to London and Sheffield. If there is any *Guinness Book of Records* for simultaneous lectureships I am sure I would have won one. I still can't think how I managed but I suppose I was working like mad, even publishing articles in refereed journals. Just as I was getting into a really stressful situation I was offered the Edge Hill permanent job to teach Afro Asian Studies and Development Geography from August 1992.

Coming from Camden to Ormskirk was a culture shock although the rural setting of Edge Hill did provide a welcome break. I quickly built The International Centre for Development and Environmental Studies (ICDES) research centre. When I first came the Edge Hill campus had a relatively good number of black and Asian students undertaking courses in Afro Asian and Race Relations Studies. With the restructuring of the polytechnics and introduction of student fees, both the Afro Asian Studies and Community and Race Relations departments closed down which reduced the intake of ethnic minority students abruptly. I moved completely to Geography taking more human geography courses and I enjoyed validating courses with third world components. With human geography colleagues like David Halsall, Paul Rodaway, Fiona Lewis (I used to address her as Mrs Moore of *Passage to India* as she always helped me out when I got into any problem) and later Sylvia Woodhead and Charles Rawding, I thought Human Geography had good coverage of subjects from Urban Geography to Historical and Heritage Geography. Regular field trips to Amsterdam and exotic third world destinations in China, India, Cuba and Morocco made the years pass quickly (Fig 10.6). Meanwhile I managed to improve the research and training profile of ICDES with a number of in-house publications, organising scholarly conferences, launching an on-line/print international refereed journal (GBER) short courses funded by the Danish Development Agency and the Government Republic of China. Recently I validated a South Asian popular culture module and produced an edited volume with writers from Edge Hill (Shakur and D'Souza, 2003). As I write this article I am also finalising another edited volume on Urban Geography (Shakur 2005). Teaching at Edge Hill helped to produce my three volumes as all of them were targeted to students. While I was validating modules I felt the students may benefit from a textbook written by their module leader. Indeed I am grateful to Human Geography students who undertook my international modules on urban geography, sustainability and cultural geography. I encouraged students to undertake dissertations on international and race issues (e.g. Aids in Africa, Critique on Rio and Johannesburg Summits, Impact of Racism in Blackburn). I managed to engage a number of students to help with the journal (some of them assuming the role of assistant editors with proficiency). Being away from the cosmopolitan cities of the UK I found it difficult to cultivate international studies in the sleepy hollow of Lancashire but teaching students (mainly from the northwest of the UK) third world studies was a gratifying and a challenging experience. I never thought I would stay at one place more than a year but I've now been at Edge Hill more than twelve years. I always considered myself as a city boy coming from a city with ten million people (Dhaka) now I seem to have settled in the suburbs of Preston (Penwortham) with hardly a few thousand inhabitants.

References

GBER (2001-2004) *Global Built Environment Review* Free International Journal on Architecture, Planning, Development and Environment, published by ICDES, Edge Hill College, Lancashire, UK
Monkhouse, F.J. (1965) *Principles of Physical Geography* (2nd edition) University of London Press.
Stamp, L D (1960) *Regional Geography*
Shakur, T, Dasgupta, N and Treloar, D (2001) *Unsustainable Development and the Cities of the Developing World,* Hegemon Press, Liverpool.
Shakur, T and D'Souza, K (eds) (2003) *Picturing South Asian Culture in English: Textual and Visual Representations*, Open House Press, Liverpool
Shakur, T. (2005) *Cities in Transition: Transforming the Global Built Environment,* Open House Press

Antics of an Administrator

June Ennis

When I joined Geography in 1995, I expected to find a load of dull geography tutors, a bit like High School. Instead I stepped into a world full of people who are passionate about geography in all its shapes and forms. I didn't realise geography was about so much more than maps.

It's amazing what you learn by typing tutors' notes and exam papers. I know much more about the world, the changing climate, why and how earthquakes happen, I've even learned some things

I never knew about Scotland, my home country. I've been privileged to see fossils first hand, not just under some museum glass box and I've learned so much about the area in which I live. I've become more environmentally friendly; my household recycles everything these days. When I take my children for walks on the beach now, I'm proud to be able to tell them much more about the coast than they learn in school.

As well as keeping the staff on their toes, reminding them it's time to lecture, I have the pleasure of booking all the exotic field trips we run. These can be anywhere from the Scottish Highlands to Llandudno, the Netherlands, Morocco, Poland and Tenerife to name but a few. The busiest times in the department are Visiting Days where you get to meet prospective students and their parents, it's a good feeling knowing you have given them a true insight into life in the department and at Edge Hill, especially when they arrive the following year ready to start a degree.

When I first started working in the department technology was not as advanced as it is nowadays and we have all had to keep up with the times, learning new software and systems. Originally the department office held only one or two box files of information and now the office is bursting at the seams with filing cabinets and files, so things have moved on, improved and got even busier, and hopefully more efficient.

The best thing for me about working here is the friendships I've made over the years, not just with the staff who work at Edge Hill, but also with the students who pass through. I've remained close friends with a few students and been to their engagement parties and children's christenings. It's nice to feel included in their lives and see them progress into geography-based careers. The social life in the Geography department is brilliant, staff and students have an annual BBQ to celebrate the end of the year and say goodbye to the third year students. We have nights out at Christmas, Easter, Half Term… well, for just about any reason really.

One thing I must mention is the department ghost. Yes, there definitely is one. Strange things happen, usually in the summer when there are no students around. The front door will bang as if someone has come into the building, but when you get up to investigate, there's no one around. The door is too heavy to blow in the wind, and several other members of staff have had the same experience. Spooky or what? All in all, I'd say Edge Hill is a fantastic place both to work and learn in. Students coming here couldn't get a better experience anywhere else.

11 Geography United

This section focuses on the integrative nature of modern geography at Edge Hill. Ann Chapman reflects on that most central of all geographers' skills: map drawing, and notes how changing technology has transformed her work. Peter Stein recounts how Edge Hill has always kept ahead of providing access for students to the most up to date satellite images, while Gerry Lucas explains the rise of geographical information sources, mapping packages and GIS made available by computers. James Curwen recounts some of his travels supporting geography research, while Sylvia Woodhead recounts a journey, from physical to human geography, explaining why the flat Netherlands landscape is so fascinating. Ann Worsley is fascinated by explorations: her researches took her to Papua New Guinea, living with tribes. A more recent twist sees her studying prehistoric footprints preserved on the nearby Formby shore. Finally two current research students explain aspects of their work. Vanessa Holden explains why her work on the Ribble marshes is so important, while Ann Power introduces her research linking human health to records of industrial pollution, looking at both National Health records and lake mud samples: a truly integrative geographical investigation.

A very peculiar place…

Ann Chapman

I began work at Edge Hill College in January 1988 as a temporary cartographic/laboratory technician. And after two days promised myself: 'I'll give it a month and then I'm off' It was like no other workplace I had known, used as I had been to the noisy easy companionship of open-plan architects' drawing offices with jokes and laughter and Friday lunchtime darts matches in the pub. My first impressions were of a quiet, very warm, utility building whose inhabitants like monks seemed to disappear into their cells for what I supposed was silent contemplation, only emerging pale and blinking into the daylight to teach. The layout was inconvenient, maps were stored in teaching rooms so access could be a problem and there was no communal space for coffee breaks so people seemed to stay in their rooms. And the students appeared like docile mice - truly a hushed and serious parallel universe. And it seemed an odd job: a combination of disparate skills, I knew that the cartographic side would be fine, I certainly could manage a map collection, however the laboratory was *terra incognita,* though I was assured that lab skills required were easy to pick up. Although the hierarchical culture did take some getting used to, everyone was very kind, I love maps, we needed the money, the job fitted around home and family life, I felt appreciated and so I stayed.

Earthlines

The Newsletter of the Department of Natural, Geographical and Applied Sciences at Edge Hill

ngas
natural geographical + applied sciences

THE GEOGRAPHY PROGRAMME at Edge Hill is continuing to develop with an emphasis on widening global knowledge and experience, so in November 2005 a band of twenty four NGAS geo students will

Hawaii 05: Fieldwork in a Pacific Hotspot

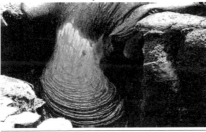

take to the air for a three-week geological field trip to Hawaii.

Visiting four of the five main islands they will study the unique collection of extinct

and active volcanoes found above this suspected, and now controversial, mantle-driven hotspot*. Gerry Lucas, who is leading the trip, expects that the visit to the Hawaiian Volcanoes National Park on Big Island will be the most explosive and interesting of the excursions, including a trek to the volcanically active fissure of Kilauea volcano which has been erupting for many years.

*You may like to visit the website at: http://www.mantleplumes.org/Hawaii.html to consider the current debate on mantle hotspots. There seems to be little evidence for their existence according to some workers, who prefer to suggest that volcanism may be linked to thinning of the crust associated with tectonic action. Hotspots and the formation of chains of volcanic islands were a substantial piece of evidence in support of the plate tectonic theory.

Gerry Lucas
lucasg@edgehill.ac.uk

Coasts at the Edge: Conference 2004

ON FRIDAY 17TH DECEMBER and for the fifth successive year, the department hosted a one-day conference which provided opportunities for students, staff and coastal professionals to come together. *Coasts at the Edge* took as its central theme how ecosystems around the margins of the Eastern Irish Sea are responding to changing physical environments and anthropogenic

activities. Guest speakers from the Proudman Oceanographic Laboratory, the RSPB and Sefton MBC, together with academic staff from Edge Hill presented their papers alongside the final year students from GEO 3003: Coastal Zone Environments and Management.

As the coasts of the United Kingdom

prepare to be highlighted by the year of the sea: 'Sea Britain 2005', through specialist publications, television and radio programmes and nationally organised events, this conference allowed for timely discussions about the past and future of our cherished coastal landscapes and environments.

Dr Annie Worsley
worsleya@edgehill.ac.uk

Jobs update: Earn enough to pay back those loans...

NOW IS BY FAR THE BEST TIME to get a geography degree because the job market for graduates is the most favourable it has ever been - a combination of technology driven

developments and EU directives has meant that there is a major shortage of geography skills - GIS and contaminated land characterisation especially. Edge Hill's new degree programmes (from 2005!) addresses these shortage areas and prepares students for this particular world of interesting work.

Check out these two sites to see the range of jobs for geo degree holders...

http://www.earthworks-jobs.com/
http://www.ends.co.uk/jobs/

Contact Gerry Lucas for more advice
lucasg@edgehill.ac.uk

issue three - autumn semester 2004

 edge hill
accredited by Lancaster University

Fig 11.1 *Earthlines* – the department newsletter

Source: NGAS

Installed in my own room, with a drawing board, tee-square, a set of lettering stencils and not much else I had to bring my own instruments from home if any cartography was to happen. Derek Mottershead convened a staff meeting and it was agreed that drawing pens would be bought for me - all very formal for what was piddling expenditure. In a previous job I had been responsible for planning, sourcing, and buying furniture and equipment for capital projects for a large local authority, with a budget to match, so this seemed amazingly penny-pinching.

Busy from the first day, my first published maps at Edge Hill were produced for a book of Paul Gamble's on *Tourism in the Third World* - first he had to submit examples of my work so his publisher could judge whether they were good enough, and then away we went, quickly establishing a good working relationship. Those were the days of tracing paper and Rotring Pens and handshading and sheets of Letraset that ran out of vowels at critical moments. After that I worked for what seemed like months on maps and diagrams intended for the second edition of 'Derek's book' actually *Environmental Systems: An introductory text* of which he was a co-author. Although exacting - the tiniest error never got past him - and very much the eminent Head of Subject, he was always good-humoured. Altogether the work was painstaking, laborious, calming and enjoyable: the cartographer in her little room churning out hand-drawn maps like miniature tapestries for erudite, courteous and (mostly) grateful colleagues.

And then an Apple Mac with an installed graphics package arrived. Not at all daunted I jumped in and taught myself computer graphics on it; lovely and intuitive to use, it was a resource shared between us all, so getting time to use it could be a problem. I remember particular polite exchanges with Fiona Lewis: 'Do you mind…?' 'No really… its all right, I'll use it after you…' as it was her machine of choice too. Later I got my very own Mac - heaven - but sadly the love affair didn't last and when the technology became outdated there was no choice: PC or nothing. Since then in cartographic terms, with patient technical help from Gerry Lucas and Peter Stein, I've graduated from Deneba Canvas to Macromedia Freehand with a little Adobe Illustrator on the side and some GIS thrown in, and from monochrome to any colour you like. I've become expert in the use and idiosyncrasies of Digimap and contributed maps and diagrams for all sorts of books, journals and PhD theses. These days there is less 'proper' cartography to do so I've 'grown the job' designing and up-dating the department website and now I'm responsible for editing *Earthlines* our biannual department newsletter, produced in full colour (Fig 11.1). I also do a couple of teaching sessions on the basics of *Digimap*. With Tas Shakur I became a publisher too – I was a partner in Open House Press: formatting, compiling the index and assisting in editing our first volume: *Picturing South Asian Culture in English*.

A major and continuing aspect of my work has been curating the department map collection. Currently we have roughly 6000 titles, from historical to current mapping and every variety of thematic maps. As technology has moved on and fewer, particularly large-scale paper maps, are routinely published, most of the collection is irreplaceable so it is a precious resource: one of Edge Hill's hidden treasures.

The drawing board has long gone to make way for more computer equipment and very early on I was delighted to hand lab work over to the more scientific hands of Kathryn Coffey. This happened for two reasons: it had quickly become apparent that there was an ever-increasing demand for cartography so the lab was often a very low priority, and without a car and dependant on a bus service that was fine in theory but incredibly unreliable in practice, travelling times to and from work effectively turned part-time into full-time hours. Something had to give. I needed more time for home and family and it was agreed that I cut six hours off my working week so that a very part-time lab technician could be appointed. To be candid, although I had experience of building science, I was also lab-illiterate, clueless and probably mildly dangerous with all those chemicals about, and I found the work repetitive and routine. In spite of Derek Mottershead's assurances that it was 'just like good-housekeeping', every time I donned the white coat I felt a fraud. I did keep a sparkling clean water-still though, so his faith in my domestic skills wasn't wildly misplaced. And it was wonderful to hand the tedium of the prescribed annual equipment inventory over to someone else.

To my continuing surprise I've worked at Edge Hill longer than I ever worked anywhere else, although there have been many 'I'll give this place another day/week/month' moments: when it is raining in Ormskirk and sunny at home, when car-parking is a pain, or when colleagues have been difficult, devious or demanding. Altogether it has been an illuminating experience and I may stay to the end…

References

Gamble, W P (1989) *Tourism and Development in Africa,* Murray, London.

White, I D Mottershead, D N and Harrison, S J (1992) *Environmental Systems: An introductory text,* Second Edition, Chapman & Hall, London.

D'Souza, Karen and Shakur, Tasleem (2003) *Picturing South Asian Culture in English: Textual and Visual Representations,* Open House Press, Liverpool.

Imaging Geography

Peter Stein

When asked by Sylvia Woodhead to pen a few words about the impact of technology on the Geography Department from a technician's point of view I was unsure of the route to take, dry statistics or a narrative.

So Max Bygraves it is. I'm going to tell a story. My personal interest in remote sensing, a modern term for satellite imagery, goes back to the early 1970s when I read an article about receiving images from an American orbiting satellite ESSA8. Building a working reception system became a goal achieved a few years later when I started receiving images from NOAA 4 on a regular basis. NOAA was the National and Oceanographic Administration and is now a component of NASA, but at the start of satellite imagery stood on its own. The USSR had a parallel service running at the same time but the secrecy imposed virtually stopped the service being openly utilised, although when it suited them imagery was released. An interesting adjunct to this is that the Russians turned their satellite off when it crossed any area considered sensitive to them but America's was always left theirs running even over home territories.

The technology to achieve this was typical of the day being a mix of mechanical photograph facsimile and the current transistor based electronics. Information on constructing such a system was initially hard to obtain from UK sources and the dull hand of government secrecy was always present, however, the Americans were open in their approach to the subject and from the outset if an interest was shown in the field of Weather Satellite Meteorology NOAA would supply you with precise details of how to build equipment and display images. It is important to mention that at this stage computers were not readily available to process images, a blend of mechanical and electronic technology had to be used. This was definitely a 'black art'.

All early NOAA satellites were classified as polar orbiters, circling the earth in an orbital sun synchronous height of 700 km and on average presented only two useable images per twelve hour cycle; they were subject to many influences that would result in poorly received images, but in 1978 everything was about to change with the entry of the European Space Agency to the field with the commissioning of Meteosat 1 in November 1977. Meteosat was in a geostationary orbit at 35000 km that gives a view of the earth with uninterrupted transmissions and therefore reliable copy. This was a huge improvement as it allowed amateurs like myself more opportunities to 'fine tune' their equipment and get better results.

In the early 1980s it was becoming easier for academic institutions to buy raw data from the Landsat series of satellites. Landsat is a system of low orbit satellites that started in 1973 continually scanning the earth in a 185km swathe transmitting the data to ground stations that process the data and present it for sale in a user friendly format corresponding to 185 x 185km squares. A library of data is available covering every part of the globe for the past twenty-five years and has become a valuable source of information. Purchased data was reprocessed locally using high end Unix based workstations and printed out on hard copy mono and colour printers.

It was at this time that I learnt of Edge Hill and the Geography Department's Landsat image processing facility. It cannot be overstated how impressive it was to have such leading edge technology and forward thinking attitudes in a local establishment. Such things were the property of red brick universities at that time.

We fast track to 1994 when I found myself on a spy mission, taking advantage of an Open Day before presenting myself for an interview for a position as network technician with the College. Coming into the Geography department on a sunny afternoon I was immediately impressed by the friendliness of those present and enthusiasm shown for geography and its allied field, particularly remote sensing, coupled to a commitment for modern computer techniques. This I had to sample.

I was appointed as network technician in October 1994 and some months later found me supporting the Geography Department as a technician. Edge Hill at this time was starting to invest heavily in modern networked personal computers. Moving away from Apple Mac and BBC influenced technologies to Windows based systems running Microsoft applications. Just to put things into perspective at this time there were only around 250 PC's throughout the whole College, currently there are over 2000.

The computer technology available to the Geography Department at the time was modest with primary image processing being undertaken on a Sun Unix workstation using bought in data. The end product was then distributed to several other windows based PC's for student use. Colour hard copy was produced from a single thin film printer, using Erdas Imagine software; this was a tedious and expensive routine. The previous description will give a flavour of the complexity of the process and the comparative user unfriendliness of the equipment.

We can now fast forward to 2004 to find that all PC workstations are Windows based with friendly graphical user interfaces and high-resolution graphics. Industry standard up-to-date software is utilised by staff and students alike and this joint approach prepares our students well for their future. Potential employers in the geographic and geological fields expect students to be competent in their use of modern software and technology. The hands on experience provided by Edge Hill gives our students an all-important advantage.

Raw data, once the domain of government is now comparatively easy to obtain from academic sources via the Internet. Packages to process the data combining both graphical and statistical information are available on Windows based operating system computers throughout the department. Hard copy, once expensive can be produced simply and economically. We supply the tools and the enthusiasm; couple it to good teaching, resulting in high quality student output.

The advance of technology over ten years is astounding. The Geography Department has committed itself to embracing these developments, investing time and expertise to ensure that the students benefit from a subject that encompasses multi-millennia dimensions. Remote sensing is a constantly evolving field that offers many and varied opportunities for study and employment.

Geo-diversity in Geography: the Rise of G20 and More

Gerry Lucas

It was April 1991 when I started at Edge Hill under the most unusual circumstances. Having worked for just one month I departed, temporarily, to undertake an eighteen-month secondment as a project geologist at the Department of Earth Sciences, Liverpool University, which had been awarded a research contract with the Department of Environment and Welsh Office to assess the sand and gravel resources of south Wales. I had raised the prospect of this secondment at my interview in December 1990 and was cut short by the chair of the panel, Professor Ruth Gee, who indicated that it was not material to the appointment process and a request for secondment would be considered, in due course, should Liverpool University be awarded the contract. They were. I requested. Edge Hill acceded. And so my first year at Edge Hill was a twilight experience, working by day in Liverpool and by occasional evening and weekend at Edge Hill, alone, preparing work for the classes I had enthusiastically agreed to cover on a part-time basis – a remote sensing course and a GIS

(Geographical Information Systems) element in the dreaded (by all) Year 3 'Geography concepts' course. It was during this time, preparing work at the weekend and evenings, that I encountered the departmental ghost. On one occasion the locked door to G20 rattled as if someone was trying to get in – yet the building was empty and totally locked. On the second occasion I was in the gents and heard a substantial coughing fit apparently coming from the prep room area – again on somewhat tremulous investigation the building was found to be locked and empty. I enquired about this at the porters' lodge (pre-security days) by telephone and was simply told that the building used to be a TB ward during World War Two.

Fig 11.2 G20 the Geo-computer room in 2004

Source: Gerry Lucas

It was during this twilight existence that the Sun micro system was acquired to replace the GEMS image processing installed during Julia McMorrow's reign. This had the effect of reducing the footprint of the image processing system from a 'room half full with equipment' to a 'room now just quarter full.' At the time of writing in 2005 the current PC based system using ERDAS Imagine is about the size of a large shoebox and probably a hundred times more powerful.

It was also at this time that the Geography department's probably greatest advance of the century was made – the birth and instatement of room G20 as the GIS computer laboratory with its networked suite of computers (Fig 11.2). Not only was G20 to become the centre for *geoinformation* at Edge Hill, in itself an addition and development of the subject, but also the student learning and social hub of the department. I guess that many past students of the 1990s and early twenty-first century will have fond memories of their time in G20 exploring this new branch of geography and the evolution of a strong geo-camaraderie that distinguishes the department above and beyond many others in the sector. It was also the time that I suffered the onerous task of having to answer the almost constant knock on the door responding to the barrage of frustrated questions such as – 'how does this…?' and 'this is not working…' and 'why won't…?' As time progressed, and IT literacy advanced, hardware became more reliable, and software easier to use, the torrent subsided. On my notice board I still have the sign I used to pin on my door when I needed sanctuary, or counselling, from the question onslaught – 'Do Not Disturb at All.' Very few took any notice. Was I too much of a nice guy?

Fig 11.3 Fieldwork in the Outer Hebrides

Source: Gerry Lucas

We are now into the fifth generation of computers in G20 with scanners, printers and our own (two) servers to distribute the image processing software (ERDAS Imagine) and GIS. Many previous students will remember the nightmare of Canvas – the graphics package - and I can think of two past students who have progressed to substantial geocareers, one at a high level in geoinformation, who appeared at my office in tears because they *'could not do Canvas.'* The progression and diversity of our computer systems – which must rank better than many institutions I have encountered elsewhere - is in part attributable to the one of the most significant events in our relatively short history of Edge Hill geocomputing – the arrival of the part-time technician Peter Stein with his full-time contribution.

It was not just the ethereal geoinformation that kept me busy at Edge Hill, but the development of the geology programme with Steve Suggitt. Guided by the firm principle, enshrined in geological mythology, that 'the best geologist is the one who has seen the most rocks' we set out to underpin the programme by a substantial rock collection (apologies to Kathy Coffey and James Curwen who had to house the collection) and fieldwork.

My first taste of fieldwork at Edge Hill was the (cold and grey) Easter trips to the Gower coast or Northumberland. But with the development of the geology programme things changed: early on in the evolutionary process we explored Fishguard and Church Stretton, my undergraduate mapping areas, and then progressed to Bude in Cornwall, before heading north to Arran and then the Western Isles of Scotland (Fig 11.3). As minor spurs we took off to Whitby, the Lake District, Anglesey and North Yorkshire and eventually maturing (at time of writing) to the pinnacle of any geo-evolutionary progression - the Mecca of geological mapping - the Ardnamurchan peninsula. Quite early on in this phylogeny however UK based fieldwork diverged with a spur taking us further afield. Mallorca as a destination appeared early on in the mid-nineties to be supported by later volcanological diversifications to Tenerife, Almeria, and Santorini culminating in the ultimate trip to Hawai'i in November 2005.

A major thread of evolution, according to Darwin, is the increase in diversity. My time at Edge Hill has seen a rapid evolutionary geodiversity, let's hope we avoid an asteroid impact? (Alvarez et al, 1980)

Reference

Alvarez, L. W., W. Alvarez, F. Asaro, and H. V. Michel. 1980. 'Extraterrestrial cause for the Cretaceous-Tertiary extinction.' *Science* 208:1095-1108

Outside Edge

James Curwen

Starting at Edge Hill at the end of summer 1997 to work in the Geography department I never realised it would take the 'lazy E' logo and me to such interesting places. Armed with rather unreliable mini buses, I had been volunteered to drive on many occasions across the United Kingdom mostly with rain, mud and the odd hiccup (Fig 11.4).

My first series of outings would take me and the research students at the time, Gareth Thompson, Richard Jones and Fiona Mann, to Cumbria, mainly Hawes Water in search of SSSI land. Some five to eight metres below this land we would find not *Ambrosia,* food of the gods, but a rather pale brown shelly clay known as marl or a rather pungent watery substance called peat. This would usually take the strength of ten men and equipment the size of a small oil rig to extract or three rather reluctant students and me.

On one such occasion we managed to lose the Livingston corer some five metres underground and realised it would take a hole the size of an Olympic swimming pool to be able to get it back. After two hours with a spade and a bucket we gave up and headed off back to the minibus which I had parked in a farmyard, only to realise it was now for some reason well and truly stuck and would not pass through the entrance without taking the roof rack off in the process. So muddy, wet and demoralised from coring, we formed a rather flimsy think tank to get us out of the situation. The suggestion to deflate the tyres was scrapped, as we had no way of pumping them back up again and then a rather angry farmer offered us his tools so we could remove the rather large and heavy roof rack. This took sometime to take off, get the mini bus out of the farmyard and then put it back on. After much to-ing and fro-ing it was put down to a bad day and we headed off back to Edge Hill and not to a local pub to toast a job well done.

After a year or two had passed of taking students to Cumbria and Anglezarke I finally had the opportunity to go on a real fieldtrip to the Outer Hebrides. This included a fourteen-hour drive to the Isle of Harris plus two hours on a rather dodgy ferry then an idyllic journey back through Lewis and Skye. All in all the best field trip I have ever been on.

As I spent more time at Edge Hill I got the chance to work for one of the Geography Department's lecturers, Dr Tasleem Shakur, and his ICDES centre. Over two weeks during the summer I took a group of fourteen Chinese delegates, of whom only one spoke good English, across the country to various towns and cities.

Fig 11.4 Geography fieldwork male voice choir, from left to right Peter Stein, Steve Suggitt, Richard Jones, Gareth Thompson, James Curwen.

Source: Barbara Lang

The trip was to look at the architecture and layouts to see if they could incorporate this into any of their designs in China, and would take me to Scotland to visit Glasgow and Edinburgh, down the country to London, passing through York, Sheffield, Milton Keynes, Cambridge to name but a few on a rather small budget. I had to navigate to each of the destinations and make sure the two minibuses, the other one driven by the then student, Jon Stevens, got there together and on time as we usually had conferences and meetings with VIPs such as Ken Livingston. Of course nothing ever went to plan and my first obstacle was a delegate who on arrival at Edinburgh was checking into his room and realised he had left his passport in Glasgow. A quick phone call to ICDES to tell Tasleem what had happened had me panicking as I was told to be careful in case it was just a ploy by the delegate to claim asylum in the United Kingdom. A few hours later everything was fine, as a journey with the delegate and Jon back to Glasgow found the passport and a camera (one of many) that was lost along the way.

Breaking down was inevitable in the old mini bus. As I tried to keep a group of delegates from wandering off and being run over along the A19 from York upon reaching my final destination London,

I knew only a few days were left till I could head back to Edge Hill. Realising that the budget was overspent and travelling and parking round London was going to be a problem, I persuaded Tasleem that using the tube system would be good for the delegates to get them around London. Appealing to his sense to save a few pounds by not having to park the minibuses or pay us both any more, I managed to cut the trip short for Jon and myself by hiring a coach out of my own pocket as ICDES and Tasleem were apparently penniless. So we headed back home, needless to say I broke down again and spent a good four hours at a motorway service station on the M1 near Birmingham awaiting the AA.

In all working at Edge Hill Geography and travelling across the country on their behalf have been some of the most interesting and yet somewhat challenging experiences I have ever dealt with.

Deep Green?

Sylvia Woodhead

I had already lived for half a century when I was appointed to teach human geography at Edge Hill in 1995. I was impressed at this evidence of adherence to Edge Hill's Equal Opportunity Policy, and also that after nearly a lifetime of teaching geology and physical geography I should now be appointed to teach human geography. I had most recently been teaching Environmental and Leisure Management, but for me being a human geographer was a new though welcome experience. This physical to human geography evolution is not uncommon: volcanoes and earthquakes grab the immediate attention. I remember avidly reading and re-reading my big three-inch thick pink-covered *Principles of Physical Geology* by Arthur Holmes, which I still have - would that have been in the sixth form or at University? It is all so long ago. Like for many others this love of the natural world gradually evolved into a realisation that people have created their own hazards by close proximity, and that it is people who study the earth's physical features and endeavour to conserve them. So a human geographer I have become.

Perceptions of the 'East'

Mention of liking geography at school reminds me that others in this volume (notably John Cater who starts his 'novella' at primary school) have written about their school experiences. Some of my reflections, on my understanding of the geography of India, can be found in my review of M M Kaye's *The Far Pavilions* (Woodhead 2003) in which I reflect on the mental images of the sub continent generated by reading this romantic novel (Fig 11.5). Until late in my sixth form career I was unquestioningly destined to become a chemist like my father before me. Quite what made me decide to change my University application from chemistry to geography I am not sure. Perhaps it was the push of chemistry as much as the pull of geography? Was it the vision of Dr Fisher teaching chemistry in his academic gown, wafting the long sleeves through the Bunsen burner flames (much as I picture the staff at Harry Potter's school, Hogwarts. I did teach for a time at a girls' boarding school which the Bronte sisters had attended, but that really is another story, and anyway I have got ahead of myself) or was it that as girls we could not cope with the knowledge that he sang counter-tenor in the church choir. Suffice it to say that I had a lucky escape.

I knew I was going to get into Leeds University to study geography when at my interview, arriving with a cold and clutching a box of tissues (I always had a cold in those days. I also remember the frequent pea-souper fogs: no relationship, perhaps) I was shown into a room reeking of menthol where Barry Garner sat behind a desk sneezing and clutching at paper handkerchiefs. After comparing cold symptoms he asked me a few desultory questions and offered me a place. Then followed the best three years of my life. For the mathematicians among you this was the early 1960s, and Jack Straw was Leeds University Union President making far left speeches - how times have changed. At Leeds University I studied subsidiary geology, which I loved, and botany, which I did not, and though the geography course was balanced with social, economic and political geography, planning and regional geography; West Yorkshire (I once counted fifty five 'very obviously' in one of Doc Fowler's lectures) and the USSR, where Glanville Jones managed to get a reference to Wales into every lecture, I specialised in physical geography.

From Leeds I moved to Aberystwyth for three years research in coastal geomorphology. Not for me a successful or happy time. As a big city girl, I didn't adapt very well to remotest Wales. Nevertheless I studied beach processes, movement of sand and shingle on the beach from Aberdovey north past Towyn where a shingle spit has closed the small Dysynni estuary. Through physical methods of beach profiling, and echo sounding below low water, tracing experiments on sand and shingle movement, and construction of wave refraction diagrams to calculate wave energy incident on the shore, I arrived at the conclusion that the present day coastline was in a state of dynamic equilibrium, and that the position of the coast had changed to achieve this. This led me to investigate past changes in the coastline position and orientation through studies of old OS and older maps and documentary sources. Folk tales of a lost land of Cantref y Gwaelod possibly record flooding in Cardigan Bay during the Flandrian marine transgression. From Aberystwyth I went into teaching where I discovered the great pleasure in imparting a joy of studying geography to new generations of students.

Fast forward as Peter Stein said. I could tell many stories, as John Cater has done, about the interviews for jobs I have had, successful or not. I too have been to interviews when the previous day I had accepted another job. Perhaps we all get our just deserts in the end. However in 1995, with a new outfit and a new short hair cut I was appointed at Edge Hill. Then followed a steep learning curve, rapidly reading up on the all those topics I had said so glibly at interview that I could teach. I must now have read almost all the geography books in the College library, though there are now so many electronic journals available that I can't possibly read more than a tiny fraction of what geographers now publish, in their search for the holy grail of the Research Assessment Exercise.

The Edge Hill team

Like Edge Hill geography students, I have learnt so many new skills, though unlike many staff I enjoyed my induction, meeting new colleagues, and learning about the management structure. My computer skills were less than distinguished on my appointment (they aren't very much better now: I come from a technophobe generation). I think my application was probably word processed on an old Amstrad. But Edge Hill College provided a whole suite of in-house training, and I went on it all, learning e-mail, word processing, Excel, PowerPoint, exam database updates, Web page design and e-learning. I was assured early on that Mediatech would help in all sorts of ways to provide AV support for my teaching, even to make a video. I took them up on this and accompanied David Halsall and his team, including Fiona Lewis, who I had replaced, on the second year human geography field course to The Netherlands. The video recorder was large and important looking, and people thought I was from Dutch TV. I made the video, though it was a struggle to hold the camera steady in the strong winds, and the editing and getting the soundtrack synchronised took ages, yet another area I am not skilled in, but my thanks go to Ken Harrison in supporting me in this enterprise, wisely tackled or not. I still show the video each year as introduction to the Dutch field course. Other early memories of Edge Hill fieldwork, before my physical geography knowledge was beginning to wane, include going to Mallorca as a biogeographer to replace Nigel Richardson, and being allowed to accompany geology fieldwork to Pembrokeshire.

At first at Edge Hill, as part of the human geography team I shared teaching with David Halsall, Paul Rodaway and Tasleem Shakur, on modules in resource management, countryside management and heritage landscapes. Paul had just been appointed Head of Department, and although he seemed awfully young for the post to me, he was swimming with ideas and enthusiasm for new ventures, and encouragingly willing to delegate. I enjoyed for a time acting as examinations officer for the department, co-ordinating arrangements with external examiners. From minor contributions to the year 1 human geography module, I developed a new fifteen-credit Year 1 Environment and Society module, combining this when the College adopted thirty credits as the norm for Year 1 into the present Introducing Human Geographies module, which I now lead. Through the second year methods module, I was interested to learn of the new cultural turn in human geography. Interactions between people and environment, a celebration of the sense of place, seemed to me to be the essence of geography.

Paul Rodaway introduced many changes. I was able to develop a new module in European Environments, reflecting my recent research in various European countries, and also making use of the information set up by the European Environment Agency based in Copenhagen. I remember organising an interview with the Danish Environment Minister, which took place on a train from Copenhagen to Naestved where he lived, and I was staying. I was so impressed with the cleanliness and recycling initiatives in Naestved, said to be Denmark's most industrial town. I enjoyed individuals showing me their wind turbines, and their pride in the waste burning district heating plant was understandable, but it got a little difficult when people knew of my research and I had to accept so many invitations to visit their 'facility', of which they were inordinately proud. These 'facilities' turned out to be sewage works, providing tertiary treatment to specifications way above those in Britain. Each visit started with coffee and Danish pastries in immaculately clean surroundings, exceptionally well designed. There is so much we could learn from their approaches to these unsocial but essential jobs. The module had a case study exam, and although I was sad when the module was dropped due to not fitting into a later validation, I was somewhat relieved that the difficult search for suitable examination scenarios was no longer needed.

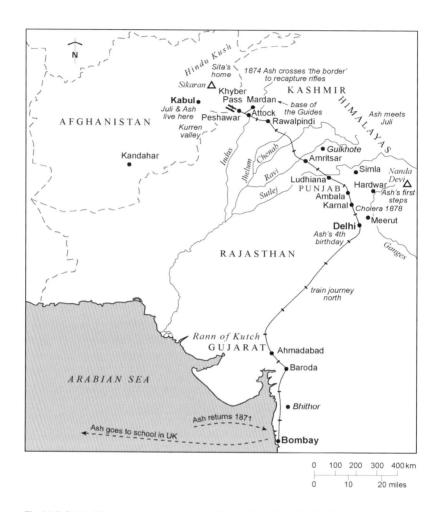

Fig 11.5 Sylvia Woodhead's mental map after reading '*The Far Pavilions*'

Map: Ann Chapman

Fieldwork has always been important to me, though one of the lows was an exceptionally wet first year field course in Langdale, with Kath Sambles and Derek Mottershead, when several days of heavy rain flooded the valley, and we were all crowded together in Elterwater Youth Hostel. I have since taken over responsibility for the Cumbria field course, and we now use the environmentally sound Castle Head field centre as a base, as so many interesting and useful environments can be reached in short distances. Students find challenging project work at Roudsea Wood NNR for physical geographers and examining Grange-over-Sands for human geographers. It is always good for students to meet different people; in Grange they meet the older age groups.

I have always had a clear focus on the importance of setting the right sort of assessment of student work. This seems crucial to me, and as a long-standing member of the College assessment sub group, I have presented papers on investigations into assessing group work and poster presentations. The geography department at Edge Hill is widely recognised for its innovative use of a great variety of assessment methods. I have developed what I hope are interesting, challenging and appropriate tasks as assessment exercises for students. These include quite a large element of formative work in Year 1. This is work which is submitted, for a mark and feedback, but which doesn't count towards the final module mark. In Year 3 I have refined a method of assessing a group discussion on a countryside management topic, typically something like the hunting debate. This reflects my own real life experience of the group interview for appointment to serve on the Cheshire and Warrington Local Access Forum, which advises on the implementation of the CROW Act 2000 to develop rights of way improvement plans. As far as evaluation goes, I have similarly designed and redefined the module questionnaires we are required increasingly to use to assess student satisfaction with our modules, though I now consider myself fortunate that my research bid to develop this evaluation work further was not successful.

Since David Halsall's retirement I have taken over responsibility for the Netherlands field course; we now fly to Amsterdam, and then use their impressive public transport. After a familiarisation to the Dutch way of life, students conduct group investigations, often of the transformations the Dutch have achieved in former run down suburbs or redundant harbour areas. I am constantly fascinated by the great differences in the Dutch approach to the environment - their love of the outdoors, even in freezing cold weather, exercise, greenery and above all water. Dutch faith in the future is also impressive, and their support for the strikingly modern designs of young architects. Most housing developments in Amsterdam include provision for self-build houses, where individuals can try out their own ideas for modern living, within some exterior constraints: Dutch individuality is mostly an indoor and hidden attribute. Impressive also is their response to Amsterdam's housing shortage. New land has been created from the former *Zuyder Zee*, simply by piling up sand until after a few years houses are built. I have been privileged to study this *Ijburg* or Water City literally rising from the sea over successive field courses. It has proved quite popular as many Amsterdammers are keen pioneers. It seems strange that the Dutch should wish to live in a place with no history whatever, but I suppose it leaves a clean palette on which to create a new and possibly ideal community. Last year I admired the tenacity of some new residents of *Ijburg* in sweeping up the blown sand out of which they were trying to create a stylish front garden. Other new suburbs on redeveloped inner harbour islands have proved attractive and safe accommodation for Amsterdam's families, and help make Amsterdam such an interesting place for geographical studies, by both staff and for student projects. In general the Dutch speak such good English that questionnaire surveys or interviews are easy to conduct. It is also a joy for geographers to see how well planned the city is. The Dutch have a clear vision for Amsterdam's future, protection of its natural environment with overall plans to achieve a sustainable city.

The Green 'Bits'

I have been interested in sustainability for some time, and in addition to research in The Netherlands, have conducted investigations in Denmark, on alternative energy, in France on natural and national park designations, and on sustainable land use in Italy. Sustainability involves restrained use of resources, thinking of the needs of both less fortunate people living in other parts of the world, and of future generations. Its basis lies in a stirring set of moral requirements for a considerate way of life. Mathis Wackernagel in California has pioneered a method of calculating an **'ecological footprint'**, a measure of the resources used by an individual, a city like London, a county like Oxfordshire or a whole country. Ecological footprint calculators are now easily accessible to all on the Internet. WWF also calculated a measure of **'earthshare'**, that is the resources available to support our existence in the world. Their conclusions are pessimistic in the extreme. Each year the earth's natural resources used (air, water, land) take more than one year to replenish. London for example is using more than eight times its 'earthshare', meaning it requires eight Earth equivalents to support the present population of London in their current lifestyle. We are literally stealing from the future. My own ecological footprint, depressingly, is much too large, derived largely from the long unsustainable journey to get me from home to Edge Hill. The ecological crisis forecast in the 1970s by Edward Goldsmith really does seem to be upon us, perhaps not in my lifetime, but certainly in the next generation.

Environmentalists call for a big change in lifestyle, but as most people dislike change in principle, this is difficult to achieve. The love affair with the car needs to be broken, but there will be such wars and fights before that is reached. Another uncertainty is to know exactly how to change - going back to a pre-industrial life of harmony in the woods, as some 'down-sizers' do might work for a minority of people, but in general we are all wedded to the advances of modern consumer products with the instant gratification of fast food, fast travel, and instantaneous communication via mobile phones and the Internet. Designer articles have become an essential part of modern life in a way my generation finds very difficult to comprehend, but even I don't want to give up modern comforts, even for the sake of the environment whose cause I supposedly espouse.

One of the troubles is that people cannot easily equate their actions and consumer purchases to effects on the environment. To help towards this I have been working on adapting a set of sustainability indicators that students and others may use to assess their actions and places of living for their level of sustainability. Recognising the value of local places is part of one of these indicators, and I have also become very interested in the geography of local places, such as the buildings and locations of Edge Hill, and hence this book. I am pleased to say that, after many presentations on Heritage Coasts at the coastal conferences organised by Ann Worsley, my suggestions that the Sefton Coast should press for designation are being followed.

Finally in 2003, for my services to geography teaching over the years, I was awarded, by the Royal Geographical Society / Institute of British Geographers, the new professional status of Chartered Geographer, one of the first cohort to be recognised in the UK. I am proud that my competence, professionalism and experience in the use of geographical knowledge and skills have been recognised in this way. I am, first and foremost, a geographer.

Reference

Woodhead, S. (2003) Picturing 'India' from M M Kaye's *The Far Pavilions*, in Shakur, T. & D'Souza, K. *Picturing South Asian Culture in English*. Open House Press, Liverpool

From Explorers to E-geography: the Adventure of a Lifetime

Annie Worsley

Early explorations

My geographical adventures began with an expedition. The mist shrouded mouth of a large, central African river clothed in dense vegetation, teeming with unseen creatures, a landscape of shifting creeks and sinuous jungle 'alleyways' led me to the shimmering, shouting waterfalls of its secret source, long sought for in a mountain hide-away amongst un-named peaks which sheltered mysterious creatures of fantasy and nightmare... at least that is the way it looked to me, aged four, from my bedroom, where my expeditionary boat constructed from bunk beds, blankets and boxes was moored. I remember the expedition as clearly as if it were yesterday though I am not absolutely sure of the name of the river. Those earliest trips into unknown territories were meticulously planned, each boat or plane specially constructed to suit the purpose of the adventure. Supplies were always carefully prepared, from ships biscuits to the armaments required to keep the crew safe.

Tales of Arabee, Samboango, the Great Red, the Congo, the Amazon, and New Guinea filled my head as I plotted and planned to lead my sister (aged two) into the unknown. Of course it was not all unknown. I had atlases and a huge, old globe that had large areas in Africa, the Americas and the East mysteriously labelled 'Empty Quarter', 'Unexplored Territory' and 'Dark Interior' and vast areas of the high seas with the label 'Sea Monsters'. I loved that globe, though over the years names faded, its sheen dulled and colours darkened. I had songs on my grandfather's knee about mysterious countries, fabulous animals, sea voyages and impenetrable mountains. But most of all I had tales about my great-great uncle, Trader Horn, who had travelled the world in search of adventure (Fig 11.6). Was geography ever more exciting? I did not know it then but it was my imaginary experiences of exploration, the maps and globe, the songs and tales of an adventuring family member that would colour and influence my whole life.

Childhood fantasies of exploration spilled out from the bedroom into 'real' expeditions through the garden and later, at the ripe old age of five, I led my first field excursion to the pond in the Big Field beyond my home. Totally out-of-bounds of course since it was bottomless and filled to the brim with monsters and my mother

Fig 11.6 Trader Horn

Source: Annie Worsley

reproached me often for asking, and asking and asking to go and fish there. Fish? No, what I really wanted was to launch a boat and capture a sea (pond) monster alive. The first expedition, then, was an illegal operation conducted in daylight. Hugely successful from the planning point of view but rather less successful in the capture of live monsters, this venture was swiftly followed by others. Indeed, so successful did I become that I grew bold and with the boldness came carelessness and arrogance and, shortly before my sixth birthday I was caught, banished to the confines of my first boat now in need of repair and reduced to despair. Despair: destroyer of souls and courage? Not for long. Banishment to my bedroom, source and resource for all my earliest expeditions, was no banishment at all. Mother could not hold me. I, scourge of the high seas, was pirate queen and leader of expeditions to the far side of the world. For a very short time I was happy but then I succumbed once more to the call of the real world.

And so, little more than twenty four hours later I was on the run, with my first mate, my four year old sister, bags packed and heading up the road for a new life. We reached the uttermost edges of the known world, the far side of the Big Field, when treachery enveloped us and we were caught and returned in imaginary chains to home and Mother. If I thought I was to be denied the life on the open road (jungle trail, high seas, desert campaign) I was wrong. This time I was allowed to explore the geography of the garden as my Mother, in her kindness (or extreme cunning?) restricted play there rather than be confined to my bedroom… In the garden I rallied my troops and together my sister and I built a jungle home with a roof made from the enormous leaves of a lush evergreen bush. The home was spacious enough for two or three (well Teddy came too) completely waterproof (those glossy leaves, I remember them so clearly.) and well supplied. However, my confidence in my ability to set up a base camp crumbled when my poor Mother came to visit the new home and realised with horror that I had stripped every single leaf from her prize Castor Oil plant. But it was a perfect roof.

Life changed then. My Mother decided that I should be taught more about plant life and she embarked on a pilgrimage of sorts. Every family holiday or trip out we tried to find out as much as possible about the plants that grew by the wayside, riverside, lakeside and mountaintop. We drew and painted the plants and the landscapes from which they came and kept diaries of what we did. We explored ancient burial sites, standing stones and chambered cairns and we pressed flowers, an act conservationists would today forbid, and painted them or their surroundings. In doing all of this however, my desire for adventure was channelled through a learning experience that rather mirrored the explorations into natural history by people in the Victorian age. The cataloguing and description of shape and form and the accounting of habitat, environment and location were fundamental to the Victorian natural historian and in many ways were crucial to the development of geography as a discipline in its own right. So too, these experiences deeply affected my perspective on the landscapes and environments I saw. And what it did for me was to prepare the ground for a lifetime's love of the British countryside, our wildest mountain and coastal landscapes especially, and a deep and abiding interest in the ancient sites where our ancestors lived and worked. I had gone from being an explorer in my imagination to a real explorer albeit confined by youth and the limits of the British Isles.

And I met Mr Savage

At school in Liverpool I indulged my passion for matters geographical and natural. The grounds of the convent school were every explorers dream, huge trees, an enormous lake, exotic shrubs and enticing paths so that even in the city the skyline was obscured by greenery. Once in secondary school (actually the same school but now with a different coloured pinafore dress and striped tie) at last I met a true savage, Mr Savage: the geographer teacher who was truly inspirational, who opened new ways of exploring the world. Geography until Mr Savage was transport, capital cities and resources. Now geography was what, why, when and how. Traditional curriculum still but served up in an intoxicating stew of questions. I was encouraged to read and find out for myself, to think and then to argue around my thinking. No other teacher had done this for me and no one else ever did until University life. If we had had computers and the Internet then I wonder where I might have ended up. But geography then developed in me a lifelong love of books. In my old school the library became the portal to new worlds, something that was needed to replace the opportunities so tantalisingly offered but denied on the wilderness-surrounded tarmac of the junior school playground. The library itself was wood panelled, highly polished and packed to the brim with volumes, some terribly old, which would be binned without ceremony in many modern schools today. But these were books from another age, of Victorian travellers and explorers, of foreign countries, tribes and landscapes that filled the shelves. And from these I developed my dual passions for geo-matters and exploration, for reading and research. It was impossible for me to imagine a world without such books. It was here, in my school library, that I found the first real reference of Trader Horn, Aloysius Smith, a Lancashire man on my mother's side of the family. Now I knew that my Grandfather's tales were true, he was an explorer and adventurer in the traditional style.

Encouraged in this way I ended up at University reading Geography in the heady, punk days of the mid-1970s, 1974–1977 to be precise, when geography was submerged in the heart of statistical analyses, systems and what to me were suffocating explanations of process, landforms, populations and city models. I believed I was born in the wrong era; that I should have been born into the time when people, men, took off in search of the unknown. Mostly I felt that I wanted to be like Freya Stark, didn't she just take off on a camel? The late great geo-explorers did not have to write essays on 'The Use of Statistics in Geography' did they? The late great geo-explorers did not have to crunch thousands of numbers to explain why rivers behaved as they did, did they? They had gone to see first hand where rivers went, why and what impact they had on people and landforms. And so I struggled with much of the first year and a half of undergraduate life by immersing myself in the literature. I was convinced that somewhere something was waiting for me but at the end of the first term in year two, I did not know what that something was.

The spring term of my second year was thrilling. I took a subsidiary course in Egyptian Archaeology followed by another in Prehistoric Archaeology, I packed the lectures in, in the hope of finding that something. Then a new member of staff arrived at Liverpool. Professor Frank Oldfield, as professor of physical geography opened up the world of change, environmental, ecological change, new (to me) methods of understanding how such changes came about and how they could be measured. I was suddenly catapulted away from statistics and slope processes into a world where the relationships between human beings and the physical environment could be explored. New methods including environmental magnetism and the use of lake sediments to reconstruct environmental history gave me the opportunity to look at the past through different eyes. I learnt about palynology, diatom analysis, sediment chemistry. I was able to combine those early loves of plants, landscapes and ancient human landscapes in my reading. I felt that my studying had some direction and meaning. Looking back I see now that my university studies mirrored the major developments in environmental change geographies.

The Frank influence

So much did Frank influence my student life that he supervised my undergraduate dissertation based on the coast. I sampled buried organic horizons in the dune slacks of Ainsdale and by using stratigraphic, pollen and diatom analyses, I was able to develop an understanding of the changes that dune systems undergo, the timescales involved and the nature of those changes. As fate would have it I have returned to that work and one aspect of my current research concentrates on understanding the evolution of the coast here in the Northwest through the analysis of buried sediments and the monitoring of contemporary ones.

Frank Oldfield was already a researcher of international pedigree with postgraduate research students, one of whom John Dearing is Professor of Geography at the University of Liverpool. The stimulus of interacting with 'postgrads' conducting research is incredibly important to the undergraduate student and, because Frank was my personal tutor and dissertation supervisor, I had regular contact with the postgraduate researchers and their work. This connection to the research activity, the specialist teaching from Frank himself had a profound influence on me. For most of the summer of 1976, a very hot summer, when I wasn't in the field, I was in his office learning to be a palynologist. Frank himself had been taught by Harry Godwin at Cambridge, who was and still is considered to be the father of British vegetation history, and so I feel honoured to be part of that line. Only recently, teaching my own students about pollen analysis, have I realised the significance of this palynological heritage. It is incredibly stimulating as an undergrad to feel that what you are doing is important and relevant and especially when that work is likely to generate new information.

It seemed to be a natural step from dissertation research to postgraduate study. Two very important factors influenced that choice. Firstly, I was captivated by research. From my very early years I wanted to ask questions, to find out about the world and postgraduate research would feed that desire. Secondly, I was a wannabe explorer. I wanted to see the world, to explore the unknown. Postgraduate research was about to provide an incredible opportunity. In the 1970s universities funded postgraduate studentships to a much greater degree and in far larger numbers than they do now. I was awarded a University of Liverpool studentship which would pay my living expenses (just) and the laboratory costs, but it would not fund the travel costs. Travel, you see, was very much on the cards, for I was bound for Papua New Guinea. Somehow I had to find the money to take me to the other side of the world. I applied to the Royal Geographical Society for help and was awarded grants from the Dudley Stamp Memorial Fund and the International Congress Fund. And then, in what I believed would be a hopeless, yet typically bold move, I applied for a scholarship at ICI. Their Brunner Mond Scholarship scheme was intended for locally (Widnes/Runcorn) born researchers in chemistry but I had nothing to lose and so I wrote a proposal about my intended fieldwork and submitted it. To my surprise, and quite honestly I am still mystified, they called me for interview along with five male chemists studying at universities across the country. Maybe it was my tale of female explorers and the opportunities of work in Papua New Guinea, maybe it was the Irish/Lancashire heritage, the 'gift-of-the-gab', maybe it was the thrill of the science, I simply do not know, but I was awarded the scholarship and so my journey to the far side of the world was assured.

Papua New Guinea

I was twenty-three when I set off around the world to Papua New Guinea. I had never flown before (incredible to believe when I have expressed how expeditions were so important to me and incredible to believe in the twenty first century when so many of our students travel so easily to Europe, the United States and beyond). Unlike modern flights today where you can fly non-stop from London to Sydney, the journey in 1979 meant flights from Manchester to London, London to Doha in the Gulf, Doha to Colombo (Sri Lanka) Colombo to Hong Kong, Hong Kong to Port Moresby, capital of Papua. And from Port Moresby, in the smallest plane I had ever seen, I flew up into the Highlands. I travelled with my PhD supervisor, Frank Oldfield and the flight from the coast took us across spectacular mountains more dramatic than any images found in books. Here were the densely vegetated and deeply incised mountains and gorges of my imagination and grandparental and parental 'knee-seat' (lap top?) tales. As we landed we were welcomed by Professor Jack Golson of the Australian National University, Canberra, an archaeologist who had worked for years in the Highlands, and taken to the Tea Research Station at the Kuk swamp.

The very name New Guinea has a romance, darkness, a mystery to it. Synonymous with head hunters, cannibals, uncharted mountains and rivers, gold and jewels, impenetrable jungle and unexplored territory New Guinea was only really opened up following the Second World War. As recently as the early 1930s maps showed an unexplored interior and only in the 1960s did maps of the Highlands show some detail at all. In the 1970s scientists from Australia and the UK began to write in earnest about the anthropology of the tribes, the medicinal properties of plants, the geology and geomorphology of the volcanic and unsettled mountain ranges. The science was interspersed with pieces about head hunting and cannibalism but it became clear that the incredibly extreme physical environment had led to the development of more than 700 different languages, a host of tribal communities who had little connection with one another and therefore a system of tribal life which hinged upon land ownership and warfare. That is, warfare using primitive (prehistoric in style) weaponry which on occasion led to the ritualistic consumption of the brains or other parts of the vanquished enemy.

By the time I arrived cannibalism had (officially) been largely outlawed in the Central Highlands though most other tribal rituals were still employed and enjoyed by the communities there. What struck me most of all was the total reliance of the people upon their physical environment: thousands of plants were used in a huge variety of ways, medicines, foodstuff, clothing, building, weaponry, decoration. Tools and weapons were made from materials provided by local vegetation and rocks. They had also developed a system of sustainable agriculture which was productive and at the same time efficient, well planned and had distinctive impacts on soil conservation. On the steepest slopes where tropical rainfall at altitude was very intense the tribal 'gardens' were able to produce significant crop yields and yet keep soil erosion to a minimum. The system of gardening was believed to have been in use for many hundreds of generations and tribal folklore told of the longevity of the practices. They had no writing, no pottery and no metal implements. In short they were modern prehistoric peoples whose equivalent in Europe and the UK had disappeared more than 6000 years ago. My research was aimed at looking at the impact of these communities upon their physical environment through palaeoecological techniques. In other words I wanted to examine the relationship of people with their tropical montane landscapes by using the methods developed by physical geographers and environmental scientists. I wanted to reconstruct the history of the people and their systems of gardening and their impact upon the environment through pollen analysis (palynology) environmental magnetism and sediment chemistry. To do that, in the absence of any written records beyond the 1950s, I needed mud.

The mud came from small volcanic crater lakes (Ipea, Egari and Pipiak) in the Highlands. Normally sediment cores are obtained from the relative safety of a boat with all the correct gear such as life jackets, together with suitably robust coring equipment (involving compressed air, tubes, a Mackereth corer and accompanying paraphernalia). It is rather difficult to get a boat, however, in the vertiginous, mountain regions of Papua New Guinea and so an alternative approach was adopted: swimming. For this expedition it meant swimming with a small corer complete with ropes, rubber bungs and no lifejackets, in water that contained a myriad of unusual organisms and which, on one occasion, was surrounded by warring, though neighbouring, tribes who were at pains (literally) to point out that the lake in question was in an area of dispute and therefore (it was entirely possible to see clearly from shore to shore) surrounded by the deepest green forest, with the shouts of the warriors and the singing of the arrows as they flew overhead. Frank Oldfield was engrossed in the muds and water and oblivious to the hostilities playing out around us and in the airs above us. But we got our cores, our tribal bodyguards earned respect from their neighbours and the lake eventually returned to the almost mythical and mystical state of peace that it had enjoyed prior to our visit. This was a defining moment for me as a geographer and it all took place in my first week. After eight days Frank had to leave for home and I was then entirely on my own apart from Wai, who acted as my bodyguard (Fig 11.7). He had helped Professor Golson in the past and 'Master Jack' had asked him to look after me. Now, the pigeon English that I had been trying to learn prior to the visit was about to become very, very useful indeed. I became 'bospela mis long graun bilong kwin,' I had a 'wasman' (guide cum bodyguard) and for almost every 'wokabaut' there was an accompanying 'ol man i stap harim wanpela i toktok' (audience). I completed more fieldwork, collecting catchment samples, mapping and measuring sites and taking photographs. But

Fig 11.7 Wai – Annie's bodyguard

Source: Annie Worsley

then I realised that if I was to really understand the relationship these communities had with their environment I needed to get much closer to them and it. I accepted an invitation from a cousin of Wai to stay with their family in one of the family tribal huts in another village. The hut was intended for use by a family member after his wedding to a girl from a neighbouring (non-warring) tribe. But the wedding had been delayed because the groom-to-be had been ill and so I became the beneficiary and was able to move in. I was not alone, however, as I shared the hut with a family of semi-domesticated pigs (mum and her babies only because the boars were too unpredictable and bad-tempered). They were by no means the only occupants of that hut but I have difficulty in describing the others since they only emerged at night with much crunching and crackling noises and with a certain degree of mobility. Often nothing could be seen with my poor human eyes unless there was a moon (though the stars were at times marvellous to see) because in the high mountains there was often cloud at night. I had, therefore, to relax, sleep and not worry about my nocturnal friends and hope that they would not find an English girl from far away at all tasty.

There were three sections to the hut: one for the pigs sleeping quarters, one for my sleeping arrangements and the central section for the fire pit which was kept glowing permanently, for warmth at night and for cooking by day. Once I had settled in to the new home I was then able to spend time with the women learning about their use of resources, plants for medicines for example, and with the men as they hunted. Most importantly I was able to learn about the system of gardening from the gardeners themselves and eventually I was able to understand why they had developed such a close, almost symbiotic relationship with this captivating, but very extreme, tropical environment. Gardens were a sustainable system of producing crops required for a variety of uses that had little impact upon soil erosion. Although the clearance and maintenance of the gardens altered the plant communities, the system required the return of key areas to grassland in the first instance and then to forest, on a cyclical basis. Areas could be left fallow for more than a generation implying familial and tribal continuity with land ownership and methods of cultivation and hence, warfare. However, the evidence from the mud (the physical geographic, environmental sedimentary evidence) showed clearly that the system of gardening utilised in the montane tropics of New Guinea did not cause soil erosion. But I was not to see this evidence until my return home.

Fig 11.8 Annie (right) on her initiation as a tribal warrior at a traditional 'sing-sing' in Papua New Guinea.

Source: Annie Worsley

To my everlasting delight I was welcomed by the village community and allowed to visit other neighbouring tribes some of whom had never seen a 'white' person. I was made an honorary warrior and 'enjoyed' some rather interesting initiation ceremonies and in that capacity was able to venture out with Wai and others to meet the neighbours. Although one visit got caught up again in tribal fighting which resulted in several injuries to the main combatants, my forays were largely peaceful. I was allowed to participate in the funeral of a 'Big Man' (a tribal chief) to 'smoke' in the men's house, taught to use a bow (not very effectively) and then to be guest of honour at a 'sing sing' (party) (Fig 11.8). These were events and activities which have totally affected my life as a woman and geographer and enormously affected the way I think about the relationship between humanity and the environment. My New Guinea experiences were in a small way, a reflection of some of the exploratory expeditions of the 19th century though I like to think that they went further in that I was able to go a little way to gaining a deeper understanding of their tribal, communal way of life which my Victorian forbears would not have been able to do.

Of course in 1979, there were no mobile phones, no GPS, and, while I was exploring and living solo with the tribe, no ordinary telephones and no way of contacting the outside world even by letter. My poor mother, who had encouraged me so much as a youngster, coped very well indeed with my absence and lack of contact (I have often thought about how she must have felt especially as my own son spent the academic year 2003/2004 in the United States with the full weight of communication technology behind him). However, when her own mother spotted an article about the white slave trade in Indonesia in a respected broadsheet, she had a few restless days. For that devastating spell my mother neither knew nor cared that Papua New Guinea was not part of Indonesia, all she was concerned about was that her daughter was at the other side of the world and incommunicado. Today field excursions are planned to the hilt, are brimming with high-tech equipment and have to be paper driven for Health and Safety and Risk Assessment, rightly so in a litigious society, but I do long for the freedom from form filling. The idea that if you want to take students to experience geography and environment first hand you can just pack up and go is long gone.

I got a call

After completing my field and laboratory work and the PhD itself I got married and took leave from matters academic and geographical to have a family. This was in the 1980s when geography, in my view, lost the plot somewhat. Examining processes, rather than trying to see connections and causal relationships appeared to be core to geography in many universities and although I was not in academic life I wondered, as I read the journals, where research such as mine would fit in to the 'new geographies' of the 1980s. Techniques such as pollen analysis too were sidelined in favour of newer technologies and some areas of geography once the preserve of the physical geographer were being lost to chemists and physicists, particularly in the realm of atmospheric and climate studies. I came to the conclusion that my days in Higher Education had all but ended with the decision to stay at home with my children.

Then, in 1993, I got a call from Frank Oldfield asking if I would like to visit the University of Liverpool. I had been several years out of full-time work having my family of four. My children were growing up and so I was beginning to look at training to become a teacher. Then I received the call up. My visit was swiftly followed by a few short-term contracts to complete some laboratory research for Frank and teach Masters students the art of pollen analysis. The work fitted very well with my young family and allowed me the chance of getting back into the swing of things, the reading, the technical skills and of course the teaching. I began to think in earnest about what I could do. I was informed by another professor of geography that it would be impossible for me as a woman and mother to get back into academic life simply because I had spent too much time 'out of it'. So I wrote, on a whim, to Nigel Richardson, an Oldfield postgraduate a few years after me. His response meant a visit to the department of geography at Edge Hill and after a long chat I happily accepted the offer of some part-time teaching. Also in post was the terrifying Steve Suggitt who had been a Liverpool postgraduate when I was a lowly undergrad all those years ago. His reputation was overturned within minutes of meeting and I immediately felt at home, happy and confident that these remarkable chaps would guide me back into academe.

For several years I pottered between Liverpool for the research and Edge Hill for the teaching, a lifestyle/workstyle that suited my busy family life. In the summer of 1998 Paul Rodaway, then the Head of Department, gave me the chance to develop and deliver the Fastrack Access Course for geography. This was a course designed to give students aged over twenty-one, the opportunity to gain access to Higher Education without having followed the more normal 'A' level route. In effect Paul was offering me a blank sheet of paper in order for me to prove myself and though that may not have been his intention I grasped the opportunity with both hands. Paul's offer was really a reflection of the department as a whole: generous, open and forward thinking; he did not know me nor what might be the implications of a 'fast track' access course to the degree programmes on offer. But I was able to look afresh at where geography was at the end of the twentieth century and revisit my own geo-learning. My lively group of enthusiastic Fastracking geographers (some of whom have become great friends and themselves gone on to geo-careers) and the sheer joy at totally owning my course made me realise what I wanted to do.

Coastal Zones

In late 1998 Paul Rodaway became Head of the School of Sciences, Sport and Technology and Nigel Richardson took over as Head of Geography. A vacancy for a physical geographer was created, I applied and was successful, surprising to me given my age and all the years out of Higher Education and so I joined a close and happy team which included Sharon Gedye (another Liverpool graduate) Gerry Lucas, Charles Rawding, Tasleem Shakur, Sylvia Woodhead and the PhD research student Richard Jones, along with Steve and Nigel and our support staff Ann Chapman, Kathryn Coffey, James Curwen, June Ennis and Peter Stein. The immediacy of their warm welcome into the fold cannot be understated and since my appointment that generosity of support has continued. My teaching included second year physical geography, most notably 'Biogeography', field courses for the first and second years and a new third year course. 'Coastal Zone Environments and Management' was another blank canvas and once again I was able to review my own work in geography, my own research such as it was and where geography itself was going. At the very eve of the twenty-first century, with geography being sidelined in secondary education, it seemed to me that there were some very serious challenges ahead and that these arose not simply out of the subject material covered in secondary schools (hugely influential on the university and subject choices made by sixth formers) but also on the way in which geography was being taught in Higher Education. We geographers have traditionally believed that our beloved discipline is very good at developing key skills in its students, skills which employers value and need. We say it is so, our professional bodies say it is so, and yet prospective students were turning away from geography in favour of the more popular, youth culture oriented subjects. A crisis developing?

I decided that the new final year coastal module could offer something new. I had a year to prepare and I therefore began the process of networking with regional coastal workers in order to gain insight into the topics and themes which most concerned them and updating my own coastal geo-knowledge. Late in 1999 I was fortunate enough to go along to an open meeting of coastal workers, land managers and professionals who were in the process of creating a new partnership out of the Sefton Coast Management Scheme which had run very successfully, with international recognition, for twenty five years. I had no agenda, was very interested in what was said, and naturally I joined in the lively discussions. Perhaps something I said caught someone's attention but a few weeks later I was invited to join the newly formed Sefton Coast Partnership as a Board member and I accepted, a little puzzled but nevertheless rather excited. Sometimes fate has a way of engineering or synchronising meetings, meetings of minds, of ideas, or of personalities, which lead to other opportunities. This was one of those times. I have been a board member ever since and my role within the Partnership has grown considerably and with it the positive knock-on effects for my students, our department and the research undertaken here.

The coastal module asks its participants to undertake a small, original piece of research, conducted in groups within a very tight schedule of lectures and fieldwork. Worse, in some eyes, it asks them to present their results and disseminate their findings to an audience of staff, students and visitors who include key coastal professionals and other visitors from outside the institution. The effect of this assessment strategy has been to successfully advocate the employability of students and for some individuals it has led to their employment in geo-related positions. Whether or not the participation at a professional event of this kind has really done the trick is yet to be evaluated however, some students have reported that the inclusion of their conference experiences within their CV and portfolios of evidence on interview day definitely raised a few eyebrows in a most positive way. Here then is some indication of that long held view that geographers are eminently employable because of the host of skills that transfer out of undergraduate life with them into the workplace. This scheme of original, independent research, meeting very strict deadlines and conference presentation is a scheme which mirrors professional activity in the real world and has become a fixed feature in our final year programme for the last five years. It would not be possible, however, without the continued support from colleagues but it clearly demonstrates how the department blossoms with the strong bonds between staff and students.

Sefton Coast

The contacts with the Sefton Coast Partnership and the professionals who support the coastal module have led to the development of a research programme which indirectly utilises some of the palaeoenvironmental skills of my PhD training all those years ago. In the same way as the opportunistic meeting with coastal workers in 1999 another 'accidental' meeting occurred in 2002 when a young woman came to the department seemingly on a whim. 'I want to do a PhD' she announced, 'er… OK', I replied. 'And I want to do it with you here at Edge Hill.' Vanessa Holden had been an undergraduate student here at Edge Hill before I had joined the team on the staff, had gone on to do a Masters and then been gainfully employed in environmental consultancy. Now, it appeared, she wanted to come back into Higher Education and do some research. We discussed several options but she was adamant she wanted coasts. Within twelve months we had secured a Research Development Fund bursary from Edge Hill and further funding from Sefton Council and Vanessa began her research career in earnest in the October of 2003. The research is distinctive. I am not a geomorphologist but I firmly believed that a deeper understanding of coastal change was necessary, especially in the light of global climate change, and a different perspective on high resolution changes could provide some clues about how and why our coasts respond to external forcing. With that in mind Vanessa began a systematic monitoring of a section of coast at Marshside (in Sefton, Merseyside) whilst at the same time utilising techniques employed in palaeoenvironmental research to characterise the sediments which she was collecting. I liked and still like the idea of applying palaeo-techniques to contemporary problems as well as developing a clearer understanding of events in the past, which is more usual. This frame of reference, evaluating the present and then looking backwards is not new, it has been used often in palynology and other related palaeo-studies, but applying it to distinctive real world problems where professionals have to try and mitigate for changes to the environment, is. This, in my view, is a significant and distinctive part of geography and of our department and it may well be where geographical enquiry in the smaller academic institutions focuses in the future. The use of a real world context I see as very effective in enhancing the employability of students but it is also very significant to the research strategies employed in departments like ours.

Another development from the coastal connections has been research work conducted with archaeologists from Merseyside who are particularly keen to expand the corpus of knowledge about the uses of coastal areas in our pre-historic past. Much has been written about the long-term evolution of our coastline here in North West England but there are missing pieces to the jigsaw. The superb work carried out by Gordon Roberts at Formby documenting and cataloguing the ephemeral human and animal footprints that appear from time to time in the sediments on the foreshore is also well known. Over the last year I have been working with Gordon and examining the micro-fossils hidden to human eyes but contained within the footprint deposits. Forensic archaeology it has been called and referred to as such by no less than the BBC in the series *Coast,* but to me it is geography through and through: setting human beings in the context of their natural environment, but doing it through the use of palaeo-techniques such as the identification of pollen, spores and tiny fragments of charcoal. This research is satisfying my long-held love of prehistory since the deposits are more than 5000 years old and also leaves me to wondering if our ancestors were in any way as close to their physical environment as the peoples of Papua New Guinea.

Since much of the funding available for research from government bodies and non-government organisations is targeted at collaborative and 'blue-skies' programmes in the United Kingdom's top (Russell group) universities, it follows that we, in the smaller (but no less important institutions in my view) need to carefully select our research activities to answer some of the questions posed by our local and regional communities and environments. This has important implications for Edge Hill and especially for geography. We need to show in the national arena that we are giving our students the very best in terms of teaching and learning experiences and that we are conducting discipline-related and pedagogic research, not only to underpin final year modules, but also to enhance our professional standing regionally, nationally and internationally. In a close knit community of scholars and students it is possible to marry excellence in teaching with excellence in research where that team is mutually supportive. I believe this to the case in our department and it explains why I am so happy there.

Back to research

For some years I had been convinced that the application of techniques used in environmental change offered the chance not only to reconstruct pollution and environmental histories but that these might also prove very useful when examined in the context of changes identified within communities, especially community health and public health histories, for example, death by certain disease types. While still a part-timer I contacted environmental health staff, epidemiologists and public health specialists in Liverpool to canvas their opinions. They were very supportive, agreeable but unable to fund any research. Some went as far as canvassing their own superiors and I have a sheaf of letters stretching over several years as testament to their interest and their agreement that this was an extremely interesting and novel approach. Applying these palaeo-techniques would provide a unique insight into health histories, they agreed, but without funding how could the work be done. As with many ideas, it often takes one individual to say 'hang on, let's try this' and so it was with Julia Miller of Halton NHS. She also agreed that the idea was important and together with her colleagues from Knowsley primary care Trust, agreed to fund a bursary. Edge Hill's RDF would also contribute to the project and at last, in October 2004 Ann Power was appointed as research student embarking on an investigation which has suddenly captivated many. The whole issue of the relationship between human health and the environment is now emerging as one of the key themes for the next Research Assessment Exercise (RAE) in 2008 and a host of other universities are looking at this as the signal to explore research along these lines. It would appear that Edge Hill now has the opportunity to be seen as an important player as these research programmes take off around the country in some of our biggest universities. It may be one way of securing some of our geo-programmes in the light of falling recruitment if our own Faculty of Health takes up the call to arms.

E-Geography

Back in 2003 Sharon Gedye left the department for the University of Plymouth to work with Professor Brian Chalkley and his team at the LTSN subject centre for Geography, Earth and Environmental Sciences. I missed her immediately as a friend and colleague but I was allowed to 'take over' her final year module 'Environmental Change'. This was a module that encapsulated everything I loved about palaeo-studies but more importantly it opened the gateway for me into 'e-learning'. Sharon had in 2001 brilliantly taken on the use of the WebCT technology, then fairly new to College and by early 2002 had devised and developed this new form of electronic delivery of her module. She had set up the course which allowed students to work through the learning 'units' remotely at times suitable to their personal circumstances, a particularly useful enterprise given the very different students of the present day. Now students need to work alongside their studies in order to support themselves financially so rather than *workhard:playhard*, students need to *workhard:workhard:playhard*. In this context the idea of remote 'e'-learning is very attractive. For 'Environmental Change' it was doubly attractive because of the degree of difficulty of the subject material. For students to pace themselves as they tackled subjects such as the use of oxygen isotopes and palaeomagnetism, seemed very sensible indeed. I took over the module with pleasure because I loved this area of specialism but also with trepidation because I am not the most competent user of computer technology. However, the module has continued to be very successful and, in the light of the growing teaching loads since Sharon's departure (she has never been replaced) may well be an important and significant part of the delivery of modules in the future. However, many students require or prefer the comfort of face-to-face contact with staff rather than remote learning but I see a great future for WebCT as part of the whole package so that students get the personal contact, can enjoy laboratory and fieldwork (central to all geographical teaching and learning) and yet remotely access information at times more suitable to their lives which, for many, include part-time jobs.

'E'-learning is heralded by many as the way forward. Access to the World Wide Web as a source of entertainment and information is almost a constant in modern life and information and communication technologies (ICT) are significant and important components of the lives of our students. Why should they not benefit from advances in electronic teaching and learning? I can see the arguments for promoting high quality learning support but I worry that the essence of geography, those connections we make with the real world through fieldwork, sensory perceptions and experiential learning could be diminished without the careful use of such technology. It is quite obviously here to stay and for now it must be utilised in a way that is truly supportive of student learning but does not open them up to the temptation of the next easy, plagiaristic download from the internet. Connection with the global geographic community is a huge advantage; had such technology been available in 1979 my mother would not have had those agonising days of worry waiting to hear that I had not been captured and sold into slavery in Indonesia. And the technological advances offered by remote sensing, GIS and graphical packages almost leave me breathless. They have served to make the world more accessible to everyone, geographers and non-geographers alike and I embrace that.

The technology now exists which could allow us to recreate the environments of long ago, the palaeo-environments of ancient Britain for example, and see how people interacted with their surroundings and what other environmental changes, missed by traditional palaeo-techniques, may have taken place but which allow us to rethink and reinterpret our past. Experimental computer technologies are looking at ways in which virtual worlds can be created but in such a way that we as human beings can interact with them. Some suggest that the ideas found in recent hi-tech Hollywood films may be in use within a decade or so allowing students to step in and experience a virtual rain forest for example. This would have enormous implications for the teaching of geography not least because it would allow the disabled to experience parts of the planet not accessible to them before. And it would allow me to 'take' my own students and colleagues into the montane forests and gardens of the Papuan tribes that I visited so many years ago.

But there is that one small part of me that remains the explorer, that almost wishes that there were areas of the world inaccessible unless on foot in person and not by satellite, package holiday or lap-top. I think that most of us who take up the 'flag' of geography are explorers too. Geographical research in the lab or field **is** exploration and now it is possible to be explorers along electronic pathways into worlds beyond our imagination, certainly beyond the imagination of that little girl in her bunk bed. Of course my selfish wish is not hindered by 'e' technologies but supported by them, it simply reflects the desire to ask questions and be curious about the world. Perhaps I am the explorer still.

Studying Sefton Sea Level Changes

Vanessa Holden

I spent three very happy years studying at Edge Hill. I made some good friends with whom I still stay in touch. The staff were so friendly and approachable I didn't hesitate to phone them up after seven years to talk to them about undertaking PhD research. They were thoroughly supportive of my wish to do the research. So now I have returned to the Geography Department to carry out research towards my PhD When I decided the time was right for me to start my research, Edge Hill was again top of my list for places to study. So now I'm back. I'm glad to say that even after a few personnel changes in the time that I spent away from Edge Hill, the same friendly and supportive atmosphere still remains. The close-knit nature of the department means that I can talk to my supervisors frequently, even just in passing for a quick five minute chat and similarly, they are able to have a greater understanding of my research. The disciplines covered by the staff are comprehensive, and I've no hesitation in approaching any of them for advice or help. Dr Annie Worsley is my Director of Studies, with Steve Suggitt as a supervisor. Dr Colin Booth from Wolverhampton University is also a supervisor, with Graham Lymbery from Sefton Metropolitan Borough Council being an advisor to the research. I'm looking at the development of the coastline in north Sefton, particularly the salt marshes of Marshside which have seen significant changes over the last 200 years.

My research study site of the north Sefton coast lies to the southern edge of the outer Ribble Estuary. Here, the salt marshes and muddy foreshore have received little attention from researchers in comparison to the sandy beaches backed by a barrier of sand dunes just a few kilometres to the south. The marshes are internationally recognised as being of major importance for their provision of habitats, being designated as a Site of Special Scientific Interest (SSSI) a Special Protection Area (SPA) and a RAMSAR site. The area has a number of landowners, predominantly Sefton Metropolitan Borough Council and English Nature, but with many other interested parties involved in the management of the area. The various bodies work together under the banner of the Sefton Coast Partnership, created in 2001 from the Sefton Coast Management Scheme (Fig 9.11).

Fig 11.9 Putting down markers in the outer Ribble estuary, from left to right Colin Booth, Ann Worsley, Vanessa Holden, Steve Suggitt.

Source: Colin Booth

The Ribble estuary is a classic funnel shaped estuary, measuring 10.75km between Southport and St. Anne's Pier and being 20km in length between Preston Dock, at its inner point stretching to Ribble Bar (Sefton MBC, 1983). The estuary has been recorded as covering between 70 and 100 km² when exposed at low water (Sefton MBC, 1983). The Marshside salt marshes are located on the southern edge of the Ribble Estuary, extending southwards towards the town of Southport, with the southernmost fringes reaching almost to Southport Pier. The location provides a strategically important site for the study of sea level changes due to its very close proximity to the British Isles isostatic hinge line (Haslett, 2000). As the line runs approximately south-west to north-east across the British Isles from the Mersey and Ribble through to the Tees Estuary in the northeast, the study site is crucially placed to assess sea level changes that are purely eustatic, reflecting a genuine sea level change.

The location also provides an important site in terms of its requirement for an integrated approach to its management, with the area having a number of demands placed upon it based on the resources that it offers. The understanding of the impacts of human activity in the area are crucial to its management, with tourism and recreation being a key economic factor in the local area, with the recreational beach associated with the town of Southport and bird watching on the marshes being major attractions for visitors and the local population. Human impacts from industry and commerce are also economically and socially important. Land reclaimed from the marshes for agriculture, sand extraction and waste disposal through the construction of the Coast Road are examples of how the location has been changed by human actions. Furthermore, issues such as pollution of both land and water are also anthropogenically induced factors that are crucial to understand to ensure integrated future strategic management. The understanding of the environmental issues surrounding the area, particularly those involving the management of wildlife and those controlled by legislation and EU Directives such as the Habitats Directive, are crucial to the integration of management strategies. The understanding of all these anthropogenic and environmental factors need to be combined with the physical processes operating in the area, such as changes in the channels in the Estuary, sea level changes and increasing vegetation.

Recent observations from Sefton Metropolitan Borough Council and from concerns expressed by local residents through various media have indicated a change in the sedimentation balance of the outer estuary, with muddy areas developing in a southerly direction towards the town of Southport's central coastal tourism attraction of the pier. This influx of mud is considered a major aesthetic issue by the local community, with potentially serious implications for the local tourism industry. The rapidity of this change coupled with the environmental, social and economic value of the salt marsh also makes it essential that investigations are undertaken into the sedimentology of this section of coastline to establish benchmark information on the processes occurring. The Ribble Estuary Shoreline Management Plan (2002) identified sea level rise and salt marsh spread as two issues within the management of the particular area of coast. Sedimentation on marshes represents a complex balance of inorganic sediment deposition and organic matter accumulation, which may occur above or below ground, and which may be subject to decomposition or compaction (Viles & Spencer, 1995). However, the responsive nature of such coastal sedimentary environments to environmental change provides an ideal location to study the causes and variations in natural surface processes. The consequences of these changes are appreciable from a number of standpoints, such as nature conservation, sea defence, navigation and industrial activity. It is imperative that because of their social, economic and environmental significance, coastal sediment dynamics are fully explored and understood, and are therefore able to be included in the strategies of coastal engineers, commercial activities, and conservation management plans (Viles & Spencer, 1995; Haslett, 2000).

A desktop survey, in conjunction with Sefton Metropolitan Borough Council coastal defence staff employed a Geographical Information System (GIS) to analyse recent aerial photographs and benchmark the position of the current coastal facies. Subsequent fieldwork has involved detailed ground verification of these aerial photographic interpretations. Further to this, a number of established and developing data collection and analytical techniques such as artificial marker horizons and environmental magnetics testing are being employed to establish an initial sedimentological database, to monitor spatial accretion rates, and to determine the historical nature of the sediment. The data collected during this fieldwork and subsequent laboratory analyses will be used with historic evidence such as aerial photographs and historic maps and charts, to determine the precise nature of coastal changes, the causes of these changes and then used as a predictive tool to anticipate future changes. This research will therefore provide a permanent record and database of the current situation and historical development of the north Sefton coast salt marshes that can be utilised to inform with regard to sea level change. Using this baseline data, other factors such as vegetation characteristics will be investigated to evaluate and model future changes, thereby contributing to the strategic planning of coastal change management. The requirement to quantify historic changes, determine current situations and subsequently predict future trends is a growing need due to accelerating changes in sea level, climate change and change in usage of estuarine environments (Lane, 2004).

The development of wetland environments is controlled by the changing balance between tidal regime, wind-wave climate, sediment supply, relative sea level and wetland vegetation (Viles and Spencer, 1995). Changes in this balance will result in future movements, further to those already brought about by previous changes in the dynamics. By assessing those changes that have already and are currently occurring and by assigning potential causal rationale to them, it is possible to identify likely future implications of changes in any of the controlling factors. Salt marshes in particular are responsive to changes in relative sea level due to their nature of being a dynamic soft sediment coastal type (Haslett, 2000).

Sefton Metropolitan Borough Council have fixed transects that are surveyed annually, and in many cases have records dating back eighty to ninety years. Seven of the thirty transects monitored by the Council have been selected to be studied as part of the research covering the coastline north of Southport into the southern part of the Ribble estuary. All fieldwork undertaken under this project is based upon these established transects. Marker horizons have been established at the study site to determine the level and subsequent rate of vertical accretion of the salt marsh. The measurement of this rate is crucial to understanding the response of the salt marsh to sea level rise, with the balance between the surface elevation of the marsh, through the process of vertical accretion, and a rise in sea level being critical (Reed, 1995; Haslett, 2000). If the rate of accretion is lower than that of local sea level rise, then the marsh will become submerged and effectively lost to the sea. If the rate of accretion is higher than that of sea level, then the marsh will develop further; and if the two levels are overall equal, then the system will reach an equilibrium (Haslett, 2000).

The marker horizons will establish a spatially orientated baseline of the rates of accretion of contemporary sediments occurring upon the surface of the marsh. Four transects are being studied, incorporating the southern edge of the salt marsh, through to the established salt marsh in the outer estuary. Thirteen sets of marker horizons have been established, with six types of horizon at each. The six types of horizon are:

- Feldspar
- Limestone
- Brick dust
- Polypropylene
- Artificial turf
- An 'undisturbed' site

The use of powdered feldspar has been well documented in academic publications. This mineral marker forms a colloidal layer in water, forming a surface upon which sediment deposition can occur. The distinct white layer is highly visible during coring procedures when sampling. Limestone dust (aggregate with an intermediate axis of less than 4mm) is a further mineral marker which has not been previously recorded as being employed in a wetland environment. The limestone will provide the same basis as the feldspar, but will provide a direct comparison for accretion rates in terms of surface texture and degradation due to bioturbation. Brick dust has similarly been used in wetland environments, but on more limited occasions. As this material is heavier than the feldspar, it may compensate for occurrences of higher energy such as may be found on the more seaward sites. The artificial turf is being studied to provide an insight into its capacity to act as a sediment trap, having characteristics associated with vegetation. As a number of types of turf are commercially available, a secondary study is being undertaken in parallel to the marker horizons to compare the differences over time between types of artificial turf with differing characteristics. Sheets of polypropylene have been buried at a known depth under the surface of the salt marsh to provide a further comparison with accretion rates recorded. The depth of marsh above the polypropylene sheeting will provide an indication of levels of accretion, or, should it be occurring, erosion. All the above marker horizons are being sampled every two months to provide a cumulative rate of accretion, whilst potentially being able to demonstrate any variations in accretion due to seasonality or tidal conditions. A further 'undisturbed' marker has been established, whereby marker pegs have been positioned in the marsh to allow intermittent recording of the depth of surface down from the top of each of the pegs across a 50cm square. This marker will provide a cautious, but direct comparison with the results of the surface and buried markers that experience disturbance during their establishment and sampling.

Samples of the contemporary surface sediment have been taken along seven transects, every fifty metres from the most landward point to the furthest accessible point towards Low Water Mark. This has produced in excess of one hundred samples for investigation. This sampling programme will be repeated every year (late summer) for up to three years. A sub-sample of the bulk sediment has been retained from each sample to undergo environmental magnetics testing. A portion of the remaining sample has been separated into specific grain size fractions, namely sand, silt and clay-sized materials. This is being done through a process of wet sieving followed by decanting. This is followed by magnetics testing on the individual size fractions.

Further sediment samples will be collected from other potential regional sources such as Formby Point, the coastline to the north of the Ribble estuary, and from the inner section of the estuary itself. Samples have also been collected from the Irish Sea bed via Proudman Oceanographic Laboratory. By carrying out magnetics measurements on these samples, it is then envisaged that it will be possible to carry out a provenance analysis of the contemporary sediments from the salt marsh surface (Yu and Oldfield, 1993; Lees and Pethick, 1995; and Walden *et al.*, 1997). From there, it is envisaged that the origin of the contemporary sediments can be identified, allowing high resolution mapping of the sediment dynamics of the coastline.

Further fieldwork will involve sediment cores being taken from each marker horizon location to establish the historical nature of the sediment deposited on the coastline over time. The data produced from this analysis will be used in conjunction with the archived data being made available by Sefton MBC to create a full picture of the development of the salt marsh with evidenced reasoning as to the current environmental setting. A Geographical Information System (GIS) will be used to analyse aerial photographs, charts and maps held by the Council. Profile lines will also be analysed, with data for the study site dating back to approximately 1913. Historical photographs will also be used to provide further evidence of the changes to the coastline. The archives also hold a substantial library of reports and both published and unpublished data which can be accessed for the purposes of this research.

Using the outcomes of the research outlined above, a highly detailed representation of the study site will be created. This will provide a permanent record against which future change can be assessed. Modelling of potential future changes is to be undertaken based upon the known baseline information obtained throughout the research, incorporating the contribution of various sediment sources, both at the present day and during the Late-Holocene. This will facilitate determination of the relative importance of changing sea level, regional and local wind and wave activity, sediment supply, climate fluctuations and anthropogenic factors on the past, present and future evolution of the coast.

References

Haslett, S.K. (2000) *Coastal Systems.* Routledge, London, 218

Lane, A. (2004) 'Bathymetric evolution of the Mersey Estuary, UK, 1906-1997: causes and effects.' *Estuarine, Coastal and Shelf Science.* 59 (2) 249-263

Lees, J.A. and Pethick, J.S. (1995) 'Problems associated with quantitative magnetic sourcing of sediments of the Scarborough to Mablethorpe coast, Northeast England, UK.' *Earth Surface Processes and Landforms.* 20, 795-806

Reed, D.J. (1995) 'The response of coastal marshes to sea level rise: Survival or submergence?' *Earth Surface Processes and Landforms.* 20, 39-48

Ribble Estuary Shoreline Management Partnership (2002) *Shoreline Management Plan.*

Sefton Metropolitan Borough Council. (1983) *Guide to the Sefton Coast Database.*

Viles, H. and Spencer, T. (1995) *Coastal Problems.* Edward Arnold Pubs. pp.350

Walden, J. Slattery, M.C. and Burt, T.P. (1997) 'Use of mineral magnetic measurements to fingerprint suspended sediment sources: approaches and techniques for data analysis.' *Journal of Hydrology.* 202, 353-372

Yu, L. and Oldfield, F. (1993) 'Quantitative sediment source ascription using magnetic measurements in a reservoir-catchment system near Nijar, S.E. Spain.' *Earth Surface Processes and Landforms.* 18, 441-454

The Environment and Human Health: Comparing Disease Patterns with Archives of Industrial Pollution obtained from Lake Sediments

Ann Power

Originally from Bristol, I ventured north to study Environmental Science at Wolverhampton University. I knew I wanted to continue researching after my time at Wolves when completing my final year dissertation. I investigated the use of tree leaves as a proxy indicator of vehicular pollution. I became very interested in the use of the environment and mineral magnetics to understand levels of atmospheric particulate pollution and the effects of these particulates on human health.

A few months after completing my degree I discovered a PhD research programme was available to study at Edge Hill, investigating the use of lake sediments to understand levels of pollution and links between the environment and health. Pollution, environment, health: perfect. I nervously attended the interview and after a few hours of sweating it out Dr Annie Worsley offered me the research studentship. So I have ventured further north, and am now six months into the research project.

The three-year study funded by Halton Primary Care Trust investigates the relationship between human health and atmospheric industrial pollution in Halton by using key characteristics of lake sediments. This is of great importance due to the detrimental health effects of particulate pollution.

There is a growing concern in Halton of the effects of pollution on health as the Halton / Widnes area of Merseyside is renowned for being a site of many industrial processes, in particular the chemical industry. This has resulted in extensive air and land pollution and issues have arisen over the potential health hazards to people living in the area. Therefore an understanding of the effects of environmental pollution on human health is essential.

The project is concerned with the effects of particulates (PM10s) released into the atmosphere by industrial emissions. PM10 is a classification of very fine particulate pollution with an aerodynamic diameter of less than 10μm, also representing the finest grains of particulate pollution PM2.5 (<2.5μm). These fine grains are

Fig 11.10 Collecting a lake sediment core: Ann Power in Alan Bedford's boat.

Source: Ann Power

particularly damaging to human health as PM10s remain suspended in the atmosphere for long periods of time and are able to penetrate deep into human lungs. As PM10s are so fine they are not filtered by the nasal tract and can be lodged in the alveoli where a reduction in pulmonary function can be caused. Many respiratory and cardiovascular diseases are therefore associated with PM10 pollution. Also only low levels of PM10 pollution are required to cause damaging health effects.

Industrial and vehicular combustion processes produce PM10 pollution. Therefore the health effects of this pollution will span back to the onset of the industrial revolution around 1840. Previous research investigating the effects of pollution on human health in Halton is temporally limited as only data of recent pollution levels that are monitored by the government are available. However, this project is unique as an archive of pollution histories spanning the last 300 years can be created with the use of lake sediments.

Fig 11.11 An urban lake which may record industrial pollution.

Source: Ann Power

Lake sediments are an invaluable source of environmental information and have been widely used to interpret environmental change and effects of human impact, dating back thousands of years. However little work has been carried out on lakes set in urban environments, which allow an examination of the effects of industrialisation (Fig 11.10).

Suitable sites for investigation are being continually researched. Potential sites need to be between 150–300 years old in order to achieve base-line data from pre-industrial times, although sites providing sediment around 50 years old will enhance the quality of the results. The lake sediments also need to provide complete sediment sequences. Management processes, in particular desilting, can disturb the stratigraphy of the sediment; therefore historical documents are being used to research the history of the sites.

Once a potential site has been identified a sediment core can be collected from the lake (Fig 11.11). This requires the use of a small boat, and coring equipment to extract a 1m core from the deepest part of the lake. A Gilsen corer is attached to a one metre plastic tube and lowered into the lake sediment. The device is then pulled back up from the water with the sediment remaining in the tube due to a vacuum being created. I then have the highly glamorous task of leaning over the side of the boat into the water to insert a bung in to the bottom of the tube to catch the sediment sample. The core sample captures an archive of sediment layers increasing with age down the core, which can then be prepared for analysis.

A combination of laboratory analyses will reconstruct the past industrial pollution history of Halton. Environmental magnetism techniques can be applied to identify levels of PM10 pollution. Magnetic analysis can identify and characterise PM10s and a distinction between industrial particulates and particles produced from vehicular emissions can be made. Spheroidal Carbonaceous Particulate analysis (SCP) will also be used as a primary indicator of PM10 pollution. SCPs are only produced by fuel combustion; therefore identifying SCPs in the sediment will help assess concentrations and characteristics of PM10s. Another analysis indicting pollution is X - Ray Fluorescence (XRF). This will present data regarding the composition of chemicals in the sediment. Trace elements such as lead, bromine, copper, nickel and zinc are indicative of the deposition of atmospheric pollution.

In order to produce a datable archive of pollution histories, radiocarbon dating techniques will construct the chronology of the sediment. This may also provide dates to which specific pollution events may be related.

This environmental interpretation of pollution levels throughout the last around 300 years can then be compared to past and present disease patterns of the Halton area. By researching health records, links between pollution characteristics and disease and mortality trends can be established. A better understanding of the links between characteristics and levels of particulate pollution with disease patterns can be reached.

This project highlights the importance of the complex relationship between humans and the environment.

Postscript

Sylvia Woodhead

I have always felt deeply attached to certain places; the local, the quirky and the idiosyncratic – perhaps why I fit in so well at Edge Hill. It was for this reason that I thought of writing this book about the places of Edge Hill and its people. It is through people, their activities and memories that places take on special meanings. Edge Hill is not what Yi-Fu Tuan would call a 'public symbol', an image of a place easily called up, such as Blackpool Tower or the many monuments in London; these are public symbols. These places have buildings of high 'imageability' as Kevin Lynch described in his 1960 text *In the City*, an image featured in tourist literature and easily recalled. Edge Hill is more like an example of Yi-Fu Tuan's other type of place, a 'field of care'; known mainly to local people, who derive meaning from the frequent contacts during their lives.

Fig 12.1 Edge Hill's main college buildings in 2005

Source: Gerry Lucas

As this volume recounts, Edge Hill's main buildings are very grand and much appreciated locally (Fig 12.1). They are of elegant design, they stand well and have been carefully maintained. There are no blots on the landscape, unless the wartime-built geography block is considered one (Fig 12.2). However the main building is in 1930s Lancashire Education Committee style - it is the spitting image of Morecambe High School, where I taught in the 1970s, and of many more Lancashire schools. It makes it 'home from home' to me, but to the authorities in London deciding on the building's worth, the DCMS (the Department of Culture, Media and Sport, who make decisions on Listed Buildings) don't consider it either old enough (1930s isn't old compared to archaeological sites) special enough (it is of a standard design, even if well done) or rare enough (there are lots of similar buildings in Lancashire). Neither is the building under any threat of demolition or drastic change. In fact it has a benevolent owner, Edge Hill College. For these reasons designation as Listed Building has twice been refused, much to the chagrin of College governors and West Lancashire District Council, who have put it forward for consideration. If then the classy stylish 1930s main College buildings cannot gain national recognition or protection, then the chances of any designation of the 1940s Army block that is the home of Edge Hill geography are even less likely. The building is not considered distinctive or outstanding enough, and was probably never intended to go beyond the war years; it is only of local importance. 'Block A' was no doubt built hurriedly with the nearest available resources (though some say the exterior brickwork is very good). No doubt meant as a temporary hospital ward its original use is now long gone. Investigations for this book reveal no clear picture of why it was built, nor establish any definite wartime use of the building, all is rumour or conjecture; there is no hard evidence. Some memories show clearly that it acted as a women's hall of residence in the immediate post war years, with a resident tutor either side of the entrance foyer checking in the girls at 10.00 each night. Life then was very basic and unattractive – small shared rooms, walls painted turquoise gloss, larger dormitories and primitive washing arrangements.

In 1959 Block A or the 'White House' may have accommodated the first intake of male students on the campus, but shortly afterwards it had begun to be used for teaching. It was an Education block during the 1960s, used for RE lectures, and then became the AV centre. Ken Harrison had his office in the building when he was appointed in 1964. He organised the CCTV pictures, which were taken in what is now the lecture theatre, of education students teaching pupils from local schools, and sent to a large room at the other end of the building where other students analysed the live pictures of their peers teaching children. Ken watched this through one-way glass, in a set up which would not now be allowed. Sometime in the 1960s 'The White House' became the geography block, though Steve Suggitt points out that history staff also had offices in the building for some time after that. It was gradually updated and modernised – not enough according to most staff and students, though its wartime origins have not been hidden. But behind the requirements for the building to support the modern teaching and research in twenty-first century geography lies the complex set of relationships the British have towards their heritage. Ironically in the new century we have become wedded to protecting the past - almost any redundant building, no matter what its merit, architectural or otherwise, will have its protagonists - societies to Save our Building spring up almost as soon as any modern development is proposed. It wasn't always so - in times past cities and smaller places like Ormskirk continually reworked their buildings. Ormskirk was more modern than some, and removed much of its former Georgian splendour, or allowed new roads and other developments to alter the balanced proportions of fine properties. Moreover in the post war years, following wartime destruction of many fine buildings, like Coventry Cathedral, modern designs were welcomed as being of clean lines, uncluttered and forward thinking. On a smaller and more personal scale, my mother welcomed every new post war initiative – artificial fibres like the non-iron nylon, melamine, Formica, Tupperware, the modern plastics, all to her represented clean modern, non labour-intensive indicators of a better life. What she abhorred was the stuffy clutter of Victorian ornaments, drab antimacassars, stuffed birds in glass cases and aspidistra plants, all of which she gladly disposed of when her mother died. How differently we value the past now, and regret the town centre clearances, the replacement of iconic buildings and road layouts with bland concrete structures, though some people even regretted the demolition of a monumentally ugly multi-storey car park. Many of the 1960's skyscraper blocks, now conceded to be a planning error, have been swept away without many tears. Fortunately Edge Hill has not been encumbered with any striking monstrosities of buildings from the 1960's.

Back to geography - what to do with its building? It never ceases to amaze me that modern British students value the past. Perhaps they are brainwashed by the cultural dominance of our heritage movement (though surely not by the late Fred Dibnah's TV pleas to protect the remaining factory chimneys he didn't demolish) but they do seem to value the past, this link to a place, and wish to know its history and have it interpreted. So there was student support for this book.

As a postscript for this volume I conducted a small survey among staff and students in geography at Edge Hill in 2004 (Appendix I). Most, though not all, disliked the external appearance of the geography block, commenting, as did the BBC Radio Lancashire reporter interviewing me for the radio that it does not fit in with the elegant 1930s proportions of the main building. It appears small (only one storey) compared to three storeys in the main building, and of standard Army design familiar from elsewhere; it has no immediately obviously heritage appeal. Yet in 2003 BBC2 TV's heritage programme *Restoration*, featuring Buildings at Risk, did include a POW

Fig 12.2 The geography building at Edge Hill in 2005

Source: Gerry Lucas

camp on a farm in County Durham. The fifty-five intact huts had been designated a Scheduled Ancient Monument, and bought by a local couple who intend to restore the huts, themed to show the relationship between the prisoners and the local community. Unfortunately this recognition and protection effectively reduces the chances of designation for other examples of wartime buildings deemed less well preserved or interesting.

Whereas the old fashioned **form** of the external appearance of the geography building created a negative impression, almost all students were universally complimentary about its inside workings (Fig 12.3). It **functions** very well as a working unit. The walls are vibrant with posters, maps and displays, but more than that it is about people, a small team of devoted staff, with an enviable open door policy. Lest this seem too much like a commercial I would remind you that I really am commenting on the results of a student survey. Almost all students said they would like to know more about the heritage of the building, its occupants, and of course the ghost. I hope that they will enjoy reading this book, and savour the reminiscences of previous occupants.

Fig 12.3 Interior view of the geography building in 2005

Source: Gerry Lucas

Despite the need for modernisation, almost none wanted the geography building knocked down - instead there were calls for its extension, somewhere to sit and chat (apparently gossip is good for the health and worker productivity – I read it in *The Guardian*). Perhaps now we can have that conservatory along the front that I have so frequently advocated, and while on the matter, what about solar panels on the roof to generate electricity for a hot drinks machine and water cooler. And while we are on the subject, surely there is room just outside for one of those domestic sized wind turbines that I saw at the Riso research station in Denmark. That would then make our commitment to alternative energy clearer.

Well, dreaming aside, I hope, like me, you have enjoyed this investigation into the past. How I wish I could visit the now destroyed Durning Road Building of the original Edge Hill in Liverpool, to admire the fine proportions of this former stately home, meet Miss Butterworth or even Miss Dora Smith, those formidable ladies who engendered a love of geography in their students so long ago. I wonder if I would have liked to join the gardening frenzy in the new Ormskirk site, as we think now of gardening as a retirement activity, not a student occupation, but how environmentally sound to grow all your own food. It is also a pity there is so little of the Bingley heritage left; the buildings are no longer educational; there is little to remind people of the traumatic removal of Edge Hill over the Pennines. The geography staff from the 1960s must have been a good bunch of people, a friendly lot with some amazing characters. There are stories that one member of staff in particular who always had a cup of tea in his office every afternoon, in a china teacup, with two digestive biscuits on a china plate with a cloth serviette, was so relaxed he sometimes had to be gently woken up before staff meetings. Those were the days. 1970s diversification brought in a new and different group of staff with research rather than teaching backgrounds and agendas, though still time to participate in dramatic productions and play cricket. Modernisation and upgrading followed, then foreign field courses. Strange to think that the old wartime geography building has always housed the latest technology, both for teaching support as the AV unit, and more recently in its satellite imagery and GIS equipment.

While some people have moved on, and geographers do get good jobs, see how many of them are in top management posts, others have been captured by the friendly atmosphere and stayed longer than they initially intended. Hopefully this book has shown that there is a strong future for Geography at Edge Hill.

Appendices

Appendix I Perceptions of Edge Hill's Geography Building in 2004

1. Are buildings important?	
Yes (55) They add to the image of a place, make a statement, give a professional look and confidence. They affect concentration, well-being & motivation (Year 3). They give a first impression, an insight into what's going on inside (Year 1).	No (1)

2a. What were your first impressions on seeing the outside of the Geography block?	
OK, small but pleasant, like a school. Reminded me of 'home', comfortable & attractive. Hidden away by the camouflage of plants, a cosy size, not too overpowering, manageable. Nice, but a bit shabby (Year 3). Looks well looked after, although it is old (Year 2). Good, appealing, OK, separate, its own space. Small, friendly, safe. Compact, military built, looked like huts I had at school, MY POOR RELATION. (Year 1).	Small, marginalised, scruffy, substandard & unimpressive compared to other areas in College (Staff). Old fashioned, dated (suppose that adds character) like a hospital, white. A shabby shed, not inviting, leaves a lot to be desired, worst on campus (Year 3). Dormitory barracks, tatty looking (Year 2). Like a hut, neglected. Bit run down, gives a negative impression (Year 1). Temporary. Awful hut, a shed, a bit scruffy (PGCE).

2b. What were your first impressions on seeing the inside of the Geography block?	
Welcoming, lively, full of energy, packed with information, samples, displays, but most importantly happy folk. Functional & effective (Staff). Bigger than it looks outside, well maintained, warm & friendly, cosy, calm. Better than outside, nice, good atmosphere. Small, very personal, names not numbers, sense of humour. Homely, everyone easy to find. Photos on display (Year 3). Clean, but dated. Modern rooms, vibrant, good facilities, homely, an environment that welcomes (Year 2). Well set out, easy to get around, busy, welcoming. Lots of displays, notices, pictures on the walls. Exciting, warm, friendly, plenty to attract, very geographical, MY FAVOURITE RELATION. (Year 1).	After a while, 'feels like home' (Staff). Old fashioned, not enough classrooms or computers, displays should be updated more, grim (Year 3). Small entrance & corridors (Year 2) Cluttered, bit dull & old, school & shed like, fairly outdated. Could do with modernisation (Year 1). Cramped, stuffy, could do with updating (PGCE).

Like the 'smallness', gives a whole team feel which is reflected in staff-student relations. Compactness gives it strength & personality, & is also reflected in the work done. Friendly, compact & 'insulated', very pleasant place to work (Staff). I like coming here. Seems like home now, community feel, very friendly environment, adds to the family character, warm, functional, a lot fitted in. Old with history behind its structure (Year 3). Cosy, family orientated (Year 2). Informative, warm & friendly atmosphere, everything we need (Year 1). OK, but quirky. Easy to find your way around (PGCE)	*Small, not enough space for group discussion work. (Year 3)*

4. Does the heritage of the geography block matter?	
Yes. It is an important part of College history & should be valued by all, not just us (Staff). Yes. It makes the building have more value when you know about its heritage, (Year 3). Yes. Atmosphere is created (Year 1).	*No (9). Feel it should be upgraded like the rest of campus, prefer a new building (Year 3)*

5. Any other comments about the geography building, past, present or future?	
I would like to see its independence retained but with considerable modernisation & upgrading (Staff). I like it. Keep it how it is. Like the way it is not shared with other subjects. Good to have everyone together. I will miss it when I graduate (Year 3). The past is very interesting and very surprising (PGCE).	*I would hate to see it go (Staff).*

Appendix II Geography building timeline

1940 — Built under the supervision of Lt. Hawthorne of Burscough Ordnance, for Military Hospital, as **Block A**

1947 — Painted white, known as **'The White House'**, softened by planting, converted as Hall, fitted with ex-WD stock.

1947-1949 — 22 student rooms, with 2 tutors, Mrs Burdett, RE & Miss Davies, Home Economics, resident either side of entrance, in rooms with fireplaces.

1953 — Education, Infant Studies (?)

1955-1957 — RE Lectures in Block A (?)

1958-1960 — Miss Redfearn, Handicrafts & 'Tottie' Holmes, Biology, resident tutors. Students had to make their own curtains.

1959 — First male students resident in the White House

1961-1962 — A B Dale was resident in a large room (now G12)

1965 — The building became the education base, with Ken Harrison as AV technician. Facilities included CCTV, dark room & editing suite for College photographer. Room 12 was equipped as a planetarium with a ceiling mounted plastic dome, which could descend so pupils could study the stars.

1967 — Education moved out, the building became geography (and history)

Individual offices for staff. Map collection in purpose made oak furniture together with drawing tables and light tables for map drawing. A Grant Projector was used to change map scales. Banda machine in office.

1975 — Two Sinclair ZX 81 computers introduced. Industry standard weather station

1980	Departmental library.
1992	NOAA satellite link,
1984	Nasua wet photocopier. Prep room and laboratories set up. First automated weather station, anemometer on roof
1985	One Apple IIE computer
1988	Xerox machine in the office. 12 geography BBC computers were housed in IT 'lab'. Cartography room with drawing board was set up, & a further geography teaching lab. Lecture theatre converted. An electric typewriter in the office
1991	New weather satellite system: Meteosat downline
1994	Personal computers/ Windows systems. ERDAS satellite image processing introduced. Geology lab converted, collection of rocks, minerals & fossils built up.
1995	Thin section lab set up. More analytical equipment
1996	Magnetics lab, research students lab, and GIS lab set up, using IDRISI and ERDAS image processing systems
1997	External steps removed & wheelchair access provided, former staff office converted to disabled toilet. Computer room redesigned, moved to end of block, more staff offices created.
2001	G20 computers upgraded
2002	ESRI's ARCGIS, the industry standard GIS package introduced
2004	G20 computers upgraded again. 2nd automatic weather station
2005	Meteosat weather system dies, not replaced; its technology now obsolete

Appendix III Geography staff at Edge Hill: a timeline

1913- ?	Miss Winchester
1915-1919	Miss Deakin
1925-1941	Miss EM Butterworth
1942 - ?	Miss Dora Smith
1950-1952	Miss DM Tovey
1951 - ?	Miss GA Williams, Mrs Muir
1957- 1982	Anna Cooper (1957-1982)
1958-1965	Mr Ivan Williams
1962-1975	Ann Smith
1964-1967	Rachel Hirst, now Bowles
1966-1992	Vic Keyte
1966-1969	Bill Marsden
1966-1971	Geoff Richardson
1973-1976	Peter Cundill
1973-1992	Paul Gamble
1973-2000	David Halsall
1976-1991	Julia Franklin, now McMorrow
1977-1979	Andrew Francis
1976-1993	Nigel Simons
1976-1994	Bob Slatter

1977–present	Steve Suggitt
1977-1996	Neil Immins
1979-1993	John Cater
1979-1995	Joan Swinhoe
1981-1982	Gregg Paget
1982-1996	Derek Mottershead (1982-1994)
1988-present	Ann Chapman
1989-present	Kathryn Coffey
1990-1991	Andrew Griffiths
1991-present	Gerry Lucas
1992-1995	Joe Howe
1992-present	Tasleem Shakur
1993-2003	Paul Rodaway (1994-1998)
1993-1996	Kath Sambles, now Collins
1993-1995	Fiona Lewis, now Young
1993-present	Nigel Richardson (1998- present)
1994-present	Peter Stein
1995–present	Sylvia Woodhead
1995-present	June Ennis
1997-2002	Sharon Gedye
1997-present	Charles Rawding

1997-2005	James Curwen
1997-present	Ann Worsley
2003-2003	Jasper Knight

Appendix IV Edge Hill Geography People

A to Z glossary of Edge Hill geography staff past and present

Rachel Bowles (née) Hirst was appointed to teach geomorphology on the new BEd degree at Edge Hill in 1964. In 1968 Rachel moved, on marriage to John Bowles, to Avery Hill College, London, now part of the University of Greenwich. She is now the Co-ordinator of the Register of Research in Primary Geography, and an Honorary Research Associate at the School of Education, University of Greenwich. She is a member of the Geographical Association's Primary and Middle Schools Section Committee and of the ICT Working Group, and was primary editor for the GA's Guidance series.

Elsie Butterworth, a former student of Edge Hill in Liverpool, taught geography at all three sites from 1925 to 1941. She became Principal in 1941 and retired in 1947.

John Cater was appointed in 1979 as lecturer in Human Geography, teaching urban and social geography, and also contributing to Community and Race Relations & Urban Policy and Race Relations degrees; he replaced Andrew Francis. In 1983 he became Head of UPRR, and in 1989 he was co-author of a major text on *'Social Geography'*. John became Head of Policy, Planning and Development in 1990, Director of Resources in 1992. John became Principal and Chief Executive of Edge Hill in 1993.

Ann Chapman, cartographer, 1988 to present, Ann designs maps for publication, manages the department map collection and designs and edits the department newsletter *Earthlines* which is published twice a year. She is a member of the Society of Cartographers and the British Cartographic Society.

Kathryn Coffey, technician 1989 to present, manages geography laboratories, preparing thin sections of rocks and managing the collection of minerals, rocks and fossils.

Anna Cooper was a historical geographer at Edge Hill from 1957 to 1982. In 1968 she was awarded a Leicester University M.A for work in British Landscapes under W.G. Hoskins. She developed the academic side of geography to a high level beyond the perceived needs of the B.Ed degree, hence the development of a BA degree (1974) at a time when geography was undergoing erosion in other institutions. Anna was Head of Geography at Edge Hill until she retired.

Peter Cundill was appointed, from Liverpool University in 1973, as lecturer in geomorphology. He taught on B.Ed and Cert.Ed courses and supervised students on teaching practice in primary, middle and secondary schools, he taught on BA course from 1974 to 76, when he started as lecturer in geography at the University of St Andrews, Fife, Scotland, and was replaced by Julia McMorrow. His research interests are in pollen analysis.

James Curwen, technician, providing student and staff support in physical geography and geology 1997 to 2005

June Ennis, departmental administrator, 1995 to present, manages the geography office and makes field trip bookings.

Andrew Francis was appointed lecturer in human geography at Edge Hill in 1977, and left in 1979 on his promotion to Senior Lecturer, Department of Town and Country Planning, Liverpool Polytechnic. Andrew is currently Chair of the Faculty Quality Committee in the Faculty of Media, Arts and Social Sciences at Liverpool John Moores University.

Paul Gamble was lecturer in tropical geography, from 1973 to 1992. Paul was head of Afro-Asian studies, and spent a year in London at the School of Oriental and African Studies. Paul died in post, and was replaced by Murray Steele.

Sharon Gedye, lecturer in physical geography 1996 to 2002, Sharon introduced a peatlands module at Edge Hill and was a strong supporter of WebCT. She moved south to become Geography Employability Officer at the Geography Earth and Environmental Sciences Subject Centre in Plymouth.

Andrew Griffiths, lecturer in Human Geography 1990 to 1992, including environmental management and statistics, he moved to Exeter University and is now Director of Wessex Trains.

David Halsall, lecturer in human geography 1973 to 2000. David came to Edge Hill from Manchester Grammar School, and taught modules on gender and industrial geography, with research in transport and the urban environment. He led fieldwork in Glasgow and The Netherlands. Since retiring David has moved to Bath.

Jamie Halsall, postgraduate research student, 2004 to present, supervised by Tasleem Shakur. Edge Hill geography undergraduate 2000-2003.

Vanessa Holden, postgraduate research student, on the development of the Ribble estuary salt marshes, supervised by Ann Worsley, Steve Suggitt and Colin Booth (University of Wolverhampton) 2003 to present. Edge Hill geography undergraduate 1992-1995.

Joe Howe taught human geography at Edge Hill 1992 to 1995, and is now senior lecturer in the School of Planning and Landscape, The University of Manchester, where his research is in water management and land use.

Neil Immins, senior lecturer in geography from 1975, Neil moved to education in 1982 and retired in 1996.

Nick James, research student in geography and Afro-Asian Studies 1996 to 2002, his PhD research was on food security in Zimbabwe. Nick now works as Associate Lecturer for the Open University.

Richard Jones, research student in physical geography 1995 to 1999, post-doctoral researcher at Liverpool University, and now lecturer in physical geography at Exeter University. Richard's research interests encompass palaeoclimatic and palaeoenvironmental reconstruction from lake sediments during the Late Glacial and Holocene. He is also interested in reconstructing past human impact on the landscape and is currently working on projects in Europe, China and Japan.

Jasper Knight, temporary lecturer in physical geography, for six months in 2003, to replace Sharon Gedye. Jasper's research is on late Pleistocene and Holocene environmental changes, particularly near glaciated coasts. He moved to Loughborough University, and is now at Exeter University.

Vivian (Vic) Keyte, lecturer in human geography 1966 to 1992, now retired and living in Dorset.

Barbara Lang, Edge Hill geography undergraduate 1994 to 1998, postgraduate research student 1998 to 2004, now working as a post doc on isotope research in the Earth Sciences department at Liverpool University.

Fiona Lewis (now Young) graduated from Manchester University, completed her PhD on the parish registers of Liverpool 1660-1750 at Liverpool University and was Lecturer in Human Geography at Edge Hill from 1993 to 1995, replacing Vic Keyte. She is now a freelance writer, with research interests in historical geography and women's issues.

Gerry Lucas, senior lecturer in geography, geology and GIS, from 1991 to present, current research interests solution micro-landforms, environmental geology, geotourism and geodiversity.

Fiona Mann, postgraduate research student, 1996 to 1999, on multi-proxy palaeo-climate reconstruction from Holocene carbonate lake sediments in North West England, is currently writing her thesis and is working in Edge Hill's library.

Bill Marsden was a Senior Lecturer in Geography at Edge Hill from 1966 to 1969. He moved on to the Education Department at the University of Liverpool in January 1970, retiring in 1997 as Emeritus Professor. During his time in Liverpool he wrote widely on geographical education and its history, and on the history of urban education.

Julia McMorrow (née Franklin) lecturer in geography at Edge Hill from 1976 to 1991, now Senior Lecturer in remote sensing, in the School of Geography at Manchester University. Her research interests include hyperspectral remote sensing of upland peat in the Pennines, and satellite remote sensing of land use cover changes in Southeast Asia. She has developed a virtual tour of the Dark Peak.

Derek Mottershead, appointed as Head of Geography in 1982, taught geomorphology, and his research was on rock weathering, on which he has published widely. He was co-author of an undergraduate text on *Environmental Systems*. Derek left Edge Hill in 1996. Derek holds Honorary Research Fellow positions at several Universities, and is now based at Portsmouth.

James Newman, postgraduate research student at Edge Hill from 1994 to 1998, supervised by Paul Rodaway. Research into gameworlds, became lecturer in Communications at Edge Hill, and is now at senior lecturer in Media and Cultural Studies at Bath Spa University, with expertise in video games and gaming culture.

Martin O'Hanlon, when Gerry Lucas was seconded to Liverpool University in 1991-1992, Martin taught his courses, then moved to ADAS in Leeds.

Gregg Paget, lecturer in geography at Edge Hill from 1981 to 82, is now senior lecturer, Department of Environmental and Geographical Sciences, MMU, with current research interests in countryside management, sustainability and multi-culturalism. He is Deputy Chairman of the National Wildflower Centre in Liverpool.

Mike Pearson, In-Service Course organiser, Woodlands Campus from 1982, Hon Sec of Ribblesdale Branch of Geographical Association from 1967 to 2002, Mike retired from fulltime lecturing, and is now a consultant in geographical education. He worked as a part time lecturer at Edge Hill, and retired again in 2002, when he was made an Honorary Member of the GA.

Ann Power, postgraduate research student. The environment and human health: comparing disease patterns with archives of industrial pollution obtained from lake sediments. 2004 - present, supervised by Ann Worsley.

Charles Rawding, appointed to lead the PGCE in geography 1997 to present. Charles also lectures in historical geography and tourism. His recent research is on the interwar period, looking at wartime archives for SW Lancashire. He is Secretary of the Central Lancashire branch of the GA.

Geoff Richardson, lecturer in geography, 1966-1972.

Nigel Richardson was appointed from Liverpool University in 1993 to lecture in physical geography, Nigel teaches modules on soil science and environmental change and is currently Head of Department.

Paul Rodaway was appointed to lecture in human geography in 1993, and is author of '*Sensuous Geographies*'. He became Head of Geography in 1994, then Dean of the School of Sciences, Sport and Technology, from 1998-2003. He is now Director of the Centre for Learning and Teaching at the University of Paisley.

Kath Sambles (now Collins) a lecturer in physical geography 1993 to 1996, with research in hydrology, instrumentation for measuring snow melt, Kath is now living in Australia.

Tasleem Shakur joined Edge Hill in 1992 as senior lecturer in Afro Asian Studies and Development Geography. Soon after his appointment Tas founded the International Centre for Development & Environmental Studies (ICDES). With the closing of the Afro Asian Studies department Tas moved completely to Geography where he later initiated an on-line journal GBER (Global Built Environment Review). Tas is currently the co-ordinator of Human Geography.

Bob Slatter, lecturer in biology from 1976 to 1994. Bob contributed to tropical geography courses & supervised weather readings

Nigel Simons was a lecturer in applied geography at Edge Hill between 1975 and 1989, contributing to the development of BA/BSc

Geography and undertaking research into coastal and estuarine management in England & Wales. He was appointed as Principal Lecturer in Geography at Lancashire Polytechnic in 1989, and became Head of the Department of Environmental Management when the Polytechnic became the University of Central Lancashire in 1992. He has since supervised and conducted research into environmental regulation and compliance, particularly amongst SMEs. Nigel returned to Edge Hill as Associate Dean in the Faculty of HMSAS in 2005.

Ann Smith, MA (Liverpool) in historical geography was a lecturer at Edge Hill from 1962. She was acting Head of Department 1976-1968 and second in command thereafter. Ann became Dean of the Faculty of Humanities in 1975, and retired in 1992.

Peter Stein, AFRRMS (Associate Fellow of the Royal Meteorological Society) AMRSS (Associate Member of the Remote Sensing Society) GIS technical support officer, 1994 to present.

Stephen Suggitt, appointed as lecturer in physical geography with a responsibility for Quaternary Studies in January 1977, now Senior Lecturer in Physical Geography and Geology, Course Leader for Geological Science and Programme Co-Ordinator for Physical Geography & Geology.

Joan Swinhoe, technician/administrator 1979-1995, typing, Xeroxing, cataloguing the geography library, keeping an equipment inventory and booking coaches for fieldtrips.

Gareth Thompson, postgraduate research student, 1997-2000 is examining the use of macrofossil, microfossil and humification records from raised peat bogs for reconstructing recent climate change. He is now at the University of Exeter and is preparing to submit his thesis.

Ivan Williams, lecturer in physical geography at Edge Hill from 1961 to 1965, with research interests in the coast of Wales.

Sylvia Woodhead, senior lecturer in human geography, 1995 to present, teaching on environmental resource management, countryside management and heritage landscape modules, research in sustainability, managing rights of way & heritage landscapes. Achieved Chartered Geographer status in 2003.

Ann Worsley, 1997 to 1999 part-time, 1999 to present as full-time, senior lecturer in physical geography, teaching environmental change, coastal environments/management, biogeography and core modules in geographical research techniques. Her research interests centre on the use of palaeo-environmental reconstruction techniques applied to a range of environments including coastal sediments (salt marsh and sand dunes) in Sefton and the man-made lakes of Merseyside and Cheshire (for records of atmospheric pollution). Achieved Chartered Geographer Status in 2005.

Chris Young, replaced Julia McMorrow for one year, and is now directing the geography programme at Christ Church University College, Canterbury. His research interests are in Quaternary Studies and coastal processes.

Appendix V Examination Papers 1948 & 2004

GEOGRAPHY SYLLABUS 1946 – 1948
TRAINING COLLEGE EXAMINATIONS BOARD
Universities of Manchester and Liverpool

Ordinary Course

The main objectives of the Ordinary Course in Geography are to ensure that the students possess:

A broad knowledge of World Geography (either general or special treatment of two or more large areas) …the natural conditions …and the human response …to be able to visualise the Earth 'as a whole made up of inter-related parts.' …

A more detailed knowledge of the geography of the British Isles studied in relation to the World background. 1. Physical conditions and 2. Human Geography

Thorough acquaintance with the use and interpretation of maps. Uses of map projections and types of map. Wherever possible, elementary surveying and practical weather observations should form part of the course of instruction.
Advanced Course

…Knowledge of… the Ordinary Course is assumed. The … students should have some opportunity of examining data at first hand and of reading standard works of an advanced character… The examination… may include a *viva voce* test. The requirements… are as follows:

Regional Geography.
The British Isles.
ONE of the following: Europe with special reference to Western Europe; the Mediterranean Lands; Monsoonal Asia; The Soviet Union; the United States of America.

Two of the following: General principles of Geography, Geomorphology, Climatology and Meteorology, Bio-geography, Historical Geography, Economic Geography, Social and Political Geography.

The examination will consist of two three-hour papers, or one of the papers will be of one and a half hours, if a thesis is offered.

TRAINING COLLEGE EXAMINATIONS BOARD
Universities of Manchester and Liverpool
PRINCIPLES OF TEACHING (EXTRACT)
SPECIAL COURSE B
June 8, 1948, 9.30 a.m. to 12.30 p.m.
(Answer five questions, of which one at least must be taken from each
section. State the ages of the children to whom your answers refer.
Special credit will be given for answers illustrated from personal
experience.)

HISTORY, GEOGRAPHY, AND SOCIAL STUDIES

12. With reference to a series of lessons on **one** topic, describe all the
maps you would use and how you would use them.

13. 'To achieve understanding of his subject the geographer, however
immature, must use the open air, the fields, the woods, the factories,
the mines and the towns.' Show how you would use such resources in
the teaching of geography in a particular school.

ART AND HANDWORK

29. 'The artist is not a special kind of man, but every man is a special
kind of artist.' Discuss this statement and its bearing on education.

30. Describe ways in which secondary schoolgirls can be trained in
appreciation of and discrimination in the choice of style and materials
used in dressmaking and home furnishings. What is the value of such
a training?

GARDENING

37. Discuss the value of gardening in the school curriculum as a
means of correlating the other subjects.

TRAINING COLLEGE EXAMINATION BOARD

Universities of Manchester and Liverpool

GEOGRAPHY II (Advanced Course)

June 17, 1948, 9.30 a.m. to 11a.m.

Edge Hill Training College, Ormskirk

(Answer **two** questions. Credit will be given for useful sketch-maps and diagrams)

Compare northern France and southern England in respect of geomorphology.

Analyse the evolution of drainage and land-forms in any river-basin well known to you.

With reference to specific examples, describe and account for the characteristics of **two** of the following types of coast: (*a*) ria; (*b*) fiord; (*c*) dune; (*d*) Dalmatian.

Discuss, with examples, the development of a superimposed drainage system.

Either, Describe the land-forms influenced by ice erosion in the highlands of Great Britain.

Or, Discuss the relationships of geomorphology with economic and social geography in any area known to you.

LANCASTER UNIVERSITY
EDGE HILL

PART II (SECOND YEAR) EXAMINATION FOR THE
DEGREE OF BACHELOR OF SCIENCE AT HONOURS
LEVEL

GEOGRAPHY FULL-TIME)

MODULE: GEO2007

TITLE: GEOMORPHOLOGY

DATE OF EXAMINATION: Friday 21 May 2004

Answer BOTH Questions in Section A (equally weighted short answer questions, 20% each) and ONE from Section B (essay, 60%).

Time allowed: 2 Hours.

Section A: Attempt BOTH questions.

What do you understand by well-sorted sediment? Compare the nature of such sediment on a high energy as opposed to a low energy beach.

Contrast the processes of weathering that might apply to granitic and basaltic igneous rocks.

Section B: Answer ONE of the following essay questions only.

Discuss the nature of landforms composed of fluvioglacial material.

Describe the main types of mass movement phenomena highlighting the links between source materials and their resulting landforms.

Discuss the processes of sediment erosion and transport within a typical braided stream channel.

With reference to specific environments write an account of the concepts of thresholds and time in landform evolution.

Index